# HYDRAULICS
# OF PIPELINES

# HYDRAULICS OF PIPELINES

## PUMPS, VALVES, CAVITATION, TRANSIENTS

**J. PAUL TULLIS**
Professor of Civil and Environmental Engineering
Utah State University

**WILEY**

A Wiley-Interscience Publication

**JOHN WILEY & SONS**

New York   Chichester   Brisbane   Toronto   Singapore

*Library of Congress Cataloging in Publication Data:*
Tullis, J. Paul.
    Hydraulics of pipelines: pumps, valves, cavitation, transients /
J. Paul Tullis.
        p.   cm.
    "A Wiley-Interscience publication."
    Bibliography:   p.
    Includes index.
    ISBN 0-471-83285-5
    1. Pipe lines—Hydrodynamics. 2. Pumping machinery. 3. Valves.
4. Cavitation. 5. Hydraulic transients. I. Title.
TJ935.T85   1989
621.8'672—dc19                                                    88-29159
                                                                      CIP

Printed in the United States of America

10 9 8 7 6 5 4 3 2

# CONTENTS

# NOMENCLATURE

| | | | |
|---|---|---|---|
| **a** | Wave speed | $C_{tv}$ | Torque coefficient |
| $A$ | Area | $C_{t\Delta p}$ | Torque coefficient |
| $A_v$ | Area of pipe at inlet to valve | $C_v$ | Flow coefficient |
| AC | Asbestos cement pipe | $cv$ | control volume |
| $B$ | Transient coefficient (Eq. 9.24) | $d,D$ | diameter |
| bhp | Brake horsepower | $d_0$ | Orifice diameter |
| C | Coefficient, pipe constraint (Eq. 8.12) | $DT$ | Time step |
| | | EGL | Energy grade line |
| $C^+, C^-$ | Compatibility equations (Eq. 9.23) | $e$ | Pipe wall thickness |
| $C1-C5$ | Constants | $E$ | Modulus of elasticity |
| $C_c$ | Contraction coefficient/ jet area/pipe area | $em$ | Motor efficiency |
| | | $ep$ | pump efficiency |
| $C_d$ | Discharge coefficient | $f$ | friction factor |
| $C_{d1}$ | Discharge coefficient | $F$ | Force |
| cfs | Cubic feet per second | gpm | Gallons per minute |
| $C_h$ | Hazen–Williams coefficient | $g$ | acceleration of gravity |
| | | $H$ | Head |
| CM | Eq. 9.29 | $H_b$ | Barometric pressure head |
| CP | Eq. 9.28 | $h_f$ | Friction loss |
| $cs$ | Control surface | $Hf$ | Friction and minor losses |
| | | $h_l$ | Minor head loss |

| | | | |
|---|---|---|---|
| $H_{max}$ | Maximum head loss | $P_b$ | Barometric pressure |
| $H_{va}$ | Absolute vapor pressure head | $P_d$ | Downstream pressure |
| HGL | Hydraulic grade line | PRV | Pressure-relief valve |
| $H_p$ | Total dynamic pump head | $P_s$ | Pumping suction pressure or surge pressure |
| $HP$ | Unknown head | | |
| hp | Horsepower | PSE | Pressure scale effects |
| $HR_1\ HR_2$ | Piezometric head of reservoir | $P_u$ | Upstream pressure |
| | | $P_v$ | Vapor pressure |
| $Ht$ | Turbine head | $P_{va}$ | Absolute vapor pressure |
| $H_{va}$ | Absolute vapor head | $P_{vg}$ | Gauge vapor pressure |
| $\Delta H$ | Head drop | $\Delta P$ | Pressure difference |
| $i$ | Interest rate | $Q$ | Flow rate |
| ID | Inside diameter | $QI$ | Initial flow |
| $k$ | Relative roughness | $QP, QPD,$ | Unknown flow rates |
| $K$ | Bulk modulus | QPT, QD, | |
| $K_l$ | minor loss coefficient | QT | |
| kW | Kilowatts | $r$ | Pipe or bubble radius |
| kW-hr | Kilowatt hours | $R$ | Hydraulic radius or radius of gyration |
| $l$ | Pipe length or characteristic length | | |
| | | $R$ | Friction coefficient (Eq. 9.25) |
| mgd | Million gallons per day | Re | Reynolds number |
| $m$ | Mass flow rate | $Rg$ | Gas constant |
| $M$ | Mass of air per unit volume | RH | Reservoir head |
| | | $s$ | Surface tension |
| $n$ | Number of years and Manning's friction factor and number of nodes in series connection | $S$ | Suction specific speed and slope of EGL |
| | | sg | Specific gravity |
| | | $SF$ | Safety factor |
| | | $t$ | Time |
| $N$ | Pump speed and pipe sections | $t_c$ | Valve closure time |
| | | $T$ | Temperature, tensile force and period of oscillation (Eq. 9.60) |
| NPSHa | Net positive suction head available | | |
| NPSHr | Net positive suction head required | $u$ | $y$-velocity component |
| | | $v$ | $x$-velocity component |
| $N_s$ | Specific speed | $V$ | Velocity |
| OD | Outside diameter | $V_0$ | Initial velocity |
| $P$ | Pressure | $V_s$ | Velocity in suction line |

| | |
|---|---|
| $v$ | Volume |
| VO | Valve opening |
| VOL | Cavity volume inside air chamber (Fig. 9.10) and in column separation |
| $w$ | $z$-velocity component |
| $W$ | Weber number or weight of fluid |
| $Wf$ | Friction energy |
| whp | Water horsepower |
| $W_p$ | Pump energy |
| $WR^2$ | Rotational inertia |
| $Wt$ | Turbine energy or external load on pipe |
| $x$ | Linear dimension |
| XL | Pipe length (Fig. 9.3) |
| XLPT | Unknown lead inside surge tank (Fig. 9.8) |
| $y$ | Linear dimension |
| $Y$ | Exponent (Eq. 6.10) |
| $z, Z$ | Elevation |
| $z_s$ | Suction lift for pump |

**Greek Symbols**

| | |
|---|---|
| $\nu$ | Kinematic viscosity |
| $\rho$ | Fluid density |
| $\mu$ | Dynamic viscosity and Poisson's ratio |
| $\gamma$ | Specific weight |
| $\lambda$ | Constant in method of characteristics |
| $\sigma$ | Cavitation index |
| $\tau$ | Wall shear stress |
| $\tau_0$ | Fully developed wall shear stress |
| $\eta$ | Pump efficiency |

**Subscripts**

| | |
|---|---|
| $i$ | Incipient |
| $cr$ | Critical |
| $id$ | Incipient damage |
| $ch$ | Chocking |
| $io, co, ido$ | Reference or known values |

# PREFACE

This book is intended to serve as a reference for practicing engineers and as a textbook for upper-division undergraduate or beginning graduate courses in hydraulic design. The text covers the hydraulic aspects of pipeline designs.

Chapter 1 is a review of basic concepts and equations of fluid flow related to design and analysis of pipelines.

Chapter 2 includes the concept of feasibility studies. The discussion addresses how decisions relating to the hydraulic design of the pipeline impact the economic analysis. Materials on pipe material, selection of pressure class of pipe, and the hydraulic design of the pipeline are included. Examples demonstrate selection of pipe diameter for gravity flow and pump systems.

Chapter 3 deals strictly with the hydraulic characteristics of centrifugal pumps and their proper selection and operation. Almost all texts discuss the internal workings of pumps and the homologous relationships. However, few reference books adequately describe the relationship between the pump and the system characteristics. Discussion and examples are given on selection of single pumps, pumps in series, and pumps in parallel. The discussion includes problems with starting, operating, and stopping pumps to avoid excessive transients. Cavitation in pumps is also addressed.

The selection of control valves for a pipeline is a subject frequently overlooked and can result in serious cavitation and other operational problems. Chapter 4 describes various types of control valves, their hydraulic characteristics, and how they interact with operation of a system. The function of values, including energy dissipation, torque, cavitation, and transients, is included.

Chapters 5 through 7 address cavitation, a subject usually treated superficially in university courses. Chapter 5 discusses the fundamentals of cavitation and the influence of cavitation on the hydraulic system. Chapter 6 gives details regarding the cavitation characteristics of valves. Examples of data are included regarding how to determine the level of cavitation for values and how to design systems to avoid excessive cavitation. Chapter 7 is a similar treatment of cavitation characteristics of orifices, elbows, and other pipeline components.

Chapters 8 and 9 introduce the concept of hydraulic transients, including the causes of transients and means for controlling them. Basic equations are used to derive a numerical method of predicting hydraulic transients.

Chapter 9 develops the method of characteristics and the finite difference solution needed for computer solution of hydraulic transients. Several boundary conditions are discussed. Emphasis is on transients caused by operation of valves. The steady-state pump boundary condition is included, but there is no discussion of pump shutdown transients.

Methods of calculating transients when column separation occurs or when air is trapped in a system are discussed in Chapter 10.

J. PAUL TULLIS

*Logan, Utah*
*January 1989*

# HYDRAULICS
# OF PIPELINES

# 1

# BASIC CONCEPTS AND EQUATIONS OF FLUID FLOW

This chapter contains a brief review of fluid properties that are important in closed conduit analysis. Next, the basic equations are reviewed and examples presented to demonstrate their application. Fluid friction, minor losses, and hydraulic and energy grade lines are discussed and demonstrated with examples. The chapter ends with a short explanation of cavitation and hydraulic transients.

## 1.1  CLASSIFICATION OF FLOW

The two general classifications of flow are laminar and turbulent. Laminar flow can occur if the liquid is highly viscous or if the velocities are very low. With laminar flow adjacent layers of liquid move without disturbing each other. If dye were continuously injected at a point, it would form a thin thread identifying a streamline. Except for molecular diffusion, the dye would not mix with the liquid. With turbulent flow, there is mixing between layers of fluid.

Viscosity of the liquid is the dominant factor for laminar flow. If viscous forces are large compared with inertial forces, the flow can be laminar. The Reynolds number, which is the ratio of inertia forces to viscous forces, is a convenient parameter for predicting if a flow condition will be laminar or turbulent. The Reynolds number Re is defined as

$$\mathrm{Re} = \rho \, \frac{Vd}{\mu} = \frac{Vd}{\nu} \qquad (1.1)$$

For pipe flow, $V$ is usually the mean velocity, $d$ diameter, $\rho$ fluid density, $\mu$ dynamic viscosity, and $\nu$ kinematic viscosity. Equation 1.1 is dimensionless, and any consistent system of units can be used.

The amount of pressure drop in a given length of pipe depends on whether the flow is laminar, turbulent, or transitioning between the two. For turbulent flow, the pressure drop varies with the length of pipe, the square of the velocity, the pipe roughness, and the density and viscosity of the fluid, and inversely with the diameter.

A friction coefficient $f$ is often used to quantify the pipe roughness. (This is discussed in detail in Section 1.4.) If $Re < 2000$, the flow in a pipe will usually be laminar. For this case, the friction coefficient $f$ is a function only of the Reynolds number and is calculated by

$$f = \frac{64}{Re} \qquad (1.2)$$

At Reynolds numbers between about 2000 and 4000 the flow is unstable due to onset of turbulence. In this range, it is impossible to determine a unique value of $f$. For $Re > 4000$, the flow becomes turbulent and $f$ is a function of both Re and pipe roughness. As Re increases, $f$ eventually becomes independent of Re and only depends on pipe roughness. This is called fully turbulent flow. The range between laminar and fully turbulent flow is a transition zone. Many practical pipe flow problems fall in this transition region.

Fluid flow can also be classified as uniform or nonuniform, steady or unsteady. Uniformity refers to spatial variations. Examples of uniform flow would be flow in a straight pipe of constant diameter or flow in a prismatic open channel at constant depth. Nonuniform flow occurs whenever there is a change of flow direction or cross section of the conduit.

Steady flow means that at any point in the system the average flow conditions do not vary with time. Unsteady flow can be subdivided according to the rate at which the velocity or pressure is varying. If changes occur slowly so that compressibility of the liquid is unimportant, the phenomenon is referred to as a *surge*. Examples would be an oscillating U-tube or the rise and fall of the water level in a surge tank.

If rapid changes occur in the velocity, pressure waves are generated and transmitted through the pipe at the acoustic velocity. Such rapid velocity changes often result in large pressure increases. This type of unsteady flow is called a hydraulic *transient* or *waterhammer*.

Piping systems must be designed for both steady and unsteady flow conditions. Chapters 1 and 2 concentrate on steady state design considerations. Unsteady flow, including both surge and transient analysis, are covered in detail in Chapters 8–10. Chapters 5–7, on cavitation, are a combination of steady and unsteady flow. Cavitation can occur in a system operating at steady flow, but the cavitation is unsteady because it is made up of traveling cavities that grow and collapse rapidly.

## 1.2  FLUID PROPERTIES

*Density* ($\rho$) is defined as mass per unit volume. It is independent of gravitational force, but does depend on temperature and pressure. For liquids, this dependence is small and can sometimes be ignored. One situation in which it cannot be ignored is during hydraulic transients. When a transient is generated by a sudden decrease in velocity, the density increases slightly, which creates an acoustic pressure wave and allows potential energy to be stored in the fluid. This is discussed in detail in Chapter 8.

For gases, the density is highly dependent on both pressure and temperature and this dependence must not be ignored. The absolute pressure $P$ and density $\rho$ are related to absolute temperature by the perfect gas law

$$P = \rho R_g T \tag{1.3}$$

in which $T$ is absolute temperature and $R_g$ is the gas constant independent of temperature and pressure. Real gases below critical pressure and above critical temperature generally obey the perfect gas law. Equation 1.3 shows that absolute pressure, density, and absolute temperature are linearly related.

*Specific weight* ($\gamma$) is a measure of the weight or gravitational force acting on a unit volume of fluid. It is related to density by

$$\gamma = \rho g \tag{1.4}$$

in which $g$ is the acceleration of gravity. Values for water at different temperatures are listed in Table 1.1.

*Pressure* ($P$) is a measure of potential energy. For a static liquid open to the atmosphere, it is directly proportional to the depth $H$ below the liquid surface. More specifically, it is the weight of water above a unit area, and can be expressed as

$$P = \rho g H \tag{1.5}$$

For a pressurized system, whether it be static or flowing, $H$ in Eq. 1.5 is the height to which fluid would rise above the given point in an open piezometer tube attached to the wall of the conduit. In the SI system, $\rho = \text{kg}/\text{m}^3$, $g = \text{m}/\text{s}^2$, $H = \text{m}$, and $P = \text{N}/\text{m}^2 = \text{Pa}$. In the English system, $H = \text{ft}$, $\rho = \text{slugs}/\text{ft}^3$, $g = \text{ft}/\text{s}^2$, $P = \text{lb}/\text{ft}^2$ and $\gamma = \text{lb}/\text{ft}^3$.

The term H is often referred to as head. More precisely it should be called pressure head or piezometric head. Pressure head can be measured with a pressure gauge installed at the point of interest or with a manometer tube. The height to which the water rises in the manometer is related to the measured gauge pressure by Eq. 1.5. Piezometric head applies when the reference datum for the pressure gauge or manometer is not at the point of

**TABLE 1.1 Properties of water in English and SI units.**

| Temp. (°F) | Specific Weight $\gamma$ (lb/ft$^3$) | Mass Density $\rho$ (slugs/ft$^3$) | Dynamic Viscosity $\mu$ (lb-s/ft$^2$) | Kinematic Viscosity $\nu$ (ft$^2$/s) | Surface Tension $s$ (lb/ft) | Vapor Pressure $P_{vA/\gamma}$ (ft.) | Bulk Modulus of Elasticity $K$ (lb/in.$^2$) |
|---|---|---|---|---|---|---|---|
| 32 | 62.42 | 1.940 | $3.746 \cdot 10^{-5}$ | $1.931 \cdot 10^{-5}$ | $0.518 \cdot 10^{-2}$ | 0.20 | $293 \cdot 10^3$ |
| 40 | 62.43 | 1.940 | 3.229 | 1.664 | 0.514 | 0.28 | 294 |
| 50 | 62.41 | 1.940 | 2.735 | 1.410 | 0.509 | 0.41 | 305 |
| 60 | 62.37 | 1.938 | 2.359 | 1.217 | 0.504 | 0.59 | 311 |
| 70 | 62.30 | 1.936 | 2.050 | 1.059 | 0.500 | 0.84 | 320 |
| 80 | 62.22 | 1.934 | 1.799 | 0.930 | 0.492 | 1.17 | 322 |
| 90 | 62.11 | 1.931 | 1.595 | 0.826 | 0.486 | 1.61 | 323 |
| 100 | 62.00 | 1.927 | 1.424 | 0.739 | 0.480 | 2.19 | 327 |
| 110 | 61.86 | 1.923 | 1.284 | 0.667 | 0.473 | 2.95 | 331 |
| 120 | 61.71 | 1.918 | 1.168 | 0.609 | 0.465 | 3.9 | 333 |
| 130 | 61.55 | 1.913 | 1.069 | 0.558 | 0.460 | 5.13 | 334 |
| 140 | 61.38 | 1.908 | 0.981 | 0.514 | 0.454 | 6.67 | 330 |
| 150 | 61.20 | 1.902 | 0.905 | 0.476 | 0.447 | 8.58 | 328 |
| 160 | 61.00 | 1.896 | 0.838 | 0.442 | 0.441 | 10.95 | 326 |
| 170 | 60.80 | 1.890 | 0.780 | 0.413 | 0.433 | 13.83 | 322 |
| 180 | 60.58 | 1.883 | 0.726 | 0.385 | 0.426 | 17.33 | 313 |
| 200 | 60.12 | 1.868 | 0.673 | 0.341 | 0.412 | 26.59 | 308 |
| 212 | 59.83 | 1.860 | 0.593 | 0.319 | 0.404 | 33.90 | 300 |

*Source: Hydraulic models*, ASCE Man. of Eng. Pract. 25, 1942.

| Temp. (°C) | N/m³ | kg/m³ | kg/m·s (centipoise) $1.792 \cdot 10^{-3}$ | m²/s (centistoke) $1.792 \cdot 10^{-6}$ | N/m $7.62 \cdot 10^{-2}$ | N/m² | Pa $204 \cdot 10^{7}$ |
|---|---|---|---|---|---|---|---|
| 0 | 9805 | 999.9 | 1.792 | 1.792 | 7.62 | 0.06 | 204 |
| 5 | 9806 | 1000.0 | 1.519 | 1.519 | 7.54 | 0.09 | 206 |
| 10 | 9803 | 999.7 | 1.308 | 1.308 | 7.48 | 0.12 | 211 |
| 15 | 9798 | 999.1 | 1.140 | 1.141 | 7.41 | 0.17 | 214 |
| 20 | 9789 | 998.2 | 1.005 | 1.007 | 7.36 | 0.25 | 220 |
| 25 | 9779 | 997.1 | 0.894 | 0.897 | 7.26 | 0.33 | 222 |
| 30 | 9767 | 995.7 | 0.801 | 0.804 | 7.18 | 0.44 | 223 |
| 35 | 9752 | 994.1 | 0.723 | 0.727 | 7.10 | 0.58 | 224 |
| 40 | 9737 | 992.2 | 0.656 | 0.661 | 7.01 | 0.76 | 227 |
| 45 | 9720 | 990.2 | 0.599 | 0.605 | 6.92 | 0.98 | 229 |
| 50 | 9697 | 988.7 | 0.549 | 0.556 | 6.82 | 1.26 | 230 |
| 55 | 9679 | 985.7 | 0.506 | 0.513 | 6.74 | 1.61 | 231 |
| 60 | 9658 | 983.2 | 0.469 | 0.477 | 6.68 | 2.03 | 228 |
| 65 | 9635 | 980.6 | 0.436 | 0.444 | 6.58 | 2.56 | 226 |
| 70 | 9600 | 977.8 | 0.406 | 0.415 | 6.50 | 3.20 | 225 |
| 75 | 9589 | 974.9 | 0.380 | 0.390 | 6.40 | 3.96 | 223 |
| 80 | 9557 | 971.8 | 0.357 | 0.367 | 6.30 | 4.86 | 221 |
| 85 | 9529 | 968.6 | 0.336 | 0.347 | 6.20 | 5.93 | 217 |
| 90 | 9499 | 965.3 | 0.317 | 0.328 | 6.12 | 7.18 | 216 |
| 95 | 9469 | 961.9 | 0.299 | 0.311 | 6.02 | 8.62 | 211 |
| 100 | 9438 | 958.4 | 0.284 | 0.296 | 5.94 | 10.33 | 207 |

measurement. These concepts will be discussed more thoroughly in subsequent sections.

Gas pressure can be measured with a pressure gauge or a liquid-filled manometer tube in the same way as for liquids. However, when applying Eq. 1.5 to relate head (or height of the liquid column in the tube) to the measured pressure, it is necessary to use $\gamma$ of the liquid in the tube, not $\gamma$ of the gas.

*Surface tension* (*s*), or more correctly surface energy, is an interesting phenomenon occurring only at interfaces. This can be a liquid–liquid, a liquid–gas, or a liquid–solid interface. It is this energy that causes a water droplet to have a somewhat spherical shape, causes liquids to rise in small tubes, and allows insects to skate on water, etc. Surface tension forming water droplets can be likened to the tension in a water filled balloon. The tension of the surface film (or of the rubber) causes the pressure inside the bubble to be greater than the external pressure. This difference in pressure $\Delta P$ is related to the surface tensions by

$$\Delta P = 2s/r \tag{1.6}$$

in which $r$ is the bubble radius. Equation 1.6 identifies the units of $s$ to be force per unit length. Surface tension is a fluid property and is only affected by temperature. (Values for water are listed in Table 1.1.) Since it is a fluid property it is always present. However, it does not always need to be included in analyses. The decision to include or ignore $s$ in fluid problems depends on its relative importance to the other forces, such as gravity or viscosity. Comparison of the relative importance of the surface-tension forces is commonly done with the Weber number $W$, which is the ratio of inertial forces to surface-tension forces defined as

$$W = \frac{V^2 L \rho}{s} \tag{1.7}$$

in which $V$ and $L$ are some characteristic velocity and length. For example, if one were interested in a gas bubble rising in a liquid column, the characteristic velocity would be the rise velocity and $L$ would be the radius of the gas bubble.

Surface tension can often be ignored if there is no free surface or interface. Therefore, for most pipe flow problems it is left out of the analysis.

*Bulk modulus of elasticity* (*K*) of a liquid is a measure of its compressibility. For many applications it is appropriate to consider liquids incompressible. In other situations, such as hydraulic transients where large pressure changes occur rapidly, compressibility is very important. The bulk modulus is defined as

$$K = -\frac{\Delta P v}{\Delta \mathbf{V}} \tag{1.8}$$

in which **V** is the liquid volume and $\Delta P$ the increase in pressure which causes a volume decrease of $\Delta$**V**. The units of $K$ are the same as for pressure (see Table 1.1). A typical value of $K$ for water at normal temperature and pressure is 300,000 psi, or 2068 mPa. To gain an appreciation for the compressibility of water, consider the pressure increase required to reduce the volume by 0.1%.

$$\Delta P = -300,000/(-0.001) = 300 \text{ psi or } 2068 \text{ kPa}$$

Even though the volume change is small for large pressure increases, it is important in the development of transient analysis. The speed of an acoustic pressure wave traveling through a rigid conduit is directly related to the bulk modulus by

$$\mathbf{a} = \sqrt{\frac{K}{\rho}} \qquad (1.9)$$

The speed **a** of the pressure wave and its magnitude are both dependent on the bulk modulus. The larger $K$ is, the greater the pressure rise during a transient. If there is a small amount of free air in the liquid, compressibility is dramatically increased, $K$ and **a** are correspondingly reduced, and transient pressures will be smaller.

*Vapor pressure* $(P_v)$ is another fluid property which is temperature dependent. A simple definition is that vapor pressure is the value of the absolute external pressure on the liquid surfaces at which boiling occurs for a given temperature. Values of $P_v$ for water are given in Table 1.1. When vaporization occurs by increasing the temperature and keeping the pressure constant, it is called boiling. Vaporization can also occur if the temperature is held constant and the pressure reduced. Called cavitation, this represents one of the serious problems relative to operation of hydraulic equipment and machinery.

For a simple example of cavitation, consider drawing water from a well by a vacuum pump. As the pressure in the suction pipe is reduced, atmospheric pressure $P_b$ acting on the water forces it into the pipe. When the suction pressure reaches vapor pressure, the water at the interface starts to vaporize and the water column stops rising. Additional suction does not further reduce the pressure. Consequently, if a pump is placed too high above the water level, it cannot supply any water.

The maximum suction lift of a pump is $(P_b - P_{va})/\gamma$. The actual placement of a pump must be significantly less than this because when water is flowing, hydraulic losses and turbulence cause the instantaneous pressure at the pump suction to be less than is calculated from the elevation difference between the water surface and the pump.

Analysis to determine if cavitation will occur involves the cavitation index, which is the ratio of forces suppressing cavitation to forces causing cavitation. Two forms of the index are

$$k_c = \frac{\Delta P}{(P - P_{vg})} \qquad (1.10a)$$

and

$$\sigma = \frac{(P - P_{vg})}{\Delta P} = \frac{(P + P_b - P_{va})}{\Delta P} \qquad (1.10b)$$

in which $P$ is a reference pressure, $P_{vg}$ the gauge vapor pressure, $P_{va}$ the absolute vapor pressure, and $P_b$ the barometric or atmospheric pressure. The force suppressing cavitation is the difference between the reference pressure and the liquid vapor pressure. The force causing cavitation is directly proportional to the local kinetic energy of the fluid which is related to $\Delta P$. The cavitation index is the ratio of these two forces. Other forms of this equation are discussed in Chapter 5.

Figure 1.1 shows graphically the relationship between $P_{vg}$ and $P_{va}$. Since vapor pressure is a fluid property, its absolute value depends only on temperature (see Table 1.1). The average atmospheric pressure at any altitude can be approximated by

$$P_b = 14.7 - 0.0005z \qquad (1.11)$$

in which $P_b$ is in psi and $z$ is in feet.

*Viscosity* is a fluid property which only has meaning when the fluid is in motion. It is a measure of a fluid's resistance to shear stresses. To understand viscosity, consider a flat plate floating on the surface of a liquid. The instant the block is placed into motion only the liquid in direct contact with the block moves. This is because the liquid molecules in contact with the solid are held by molecular attraction and will always have the same velocity as the solid. Movement of lower layers of liquid depend on shear stresses being transmitted from the upper fluid layers.

The liquid at the bottom of the container will always remain at rest. This

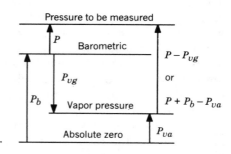

**Fig. 1.1**  Vapor pressure definitions.

creates a velocity gradient $dV/dy$ in the vertical $y$ direction. The relationship between shear stress $\tau$ and velocity gradient can be expressed as

$$\tau = \frac{\mu \, dV}{dy} \tag{1.12}$$

in which the proportionality constant $\mu$ is the dynamic viscosity.

If the solid plate in the preceding example moves at a small velocity so that the movement of the liquid is laminar, experiments have shown that a linear velocity gradient will be established and the shear stress will be equal at all elevations in the liquid for this case

$$\tau = \frac{\mu V}{y}$$

Application of the preceding principle is the basis for many instruments used to measure viscosity. Viscosity can only be measured experimentally and varies significantly with temperature. Values are listed in Table 1.1. In dealing with viscosity, the term $\mu/\rho$ is encountered frequently enough that it has been given a separate name: kinematic viscosity $\nu$.

The importance of viscosity in fluid flow problems is related to the Reynolds number Re defined by Eq. 1.1. Since Re is the ratio of inertia forces to viscous forces, smaller values of Re identify situations where viscosity is more important. The upper limit of Re above which viscous forces are not important varies with each physical problem.

Part of any hydraulic analysis is identifying which of the fluid properties are significant and must be considered. When generating design data in the laboratory it is often not physically possible to account for all important fluid properties. In such cases, the experimental data will only accurately represent the phenomenon over the range of test conditions. If the data are extrapolated to a different velocity, size, water temperature, etc., errors will be introduced due to the experimental limitations. This is referred to as a scale effect. Such scale effects will be considered in subsequent sections.

## 1.3    BASIC EQUATIONS

Solution to most fluid flow problems generally involves the application of one or more of the three basic equations: continuity, momentum, and energy. These three basic tools are developed from the law of conservation of mass, Newton's second law of motion, and the first law of thermodynamics.

**Continuity Equation**

The simplest form of this equation is for one dimensional incompressible steady flow in a closed conduit. Applying continuity between any two sections gives

$$A_1 V_1 = A_2 V_2 = Q \tag{1.13}$$

in which $A$ is the cross sectional area of the pipe, $V$ the mean velocity at that same location, and $Q$ the flow rate. Equation 1.13 is valid for any rigid conduit as long as there is no addition or loss of liquid between the sections. If the fluid is compressible, density must be added to the equation

$$\rho_1 A_1 V_1 = \rho_2 A_2 V_2 = m \tag{1.14}$$

in which $m$ is the mass flow rate.

For two- or three-dimensional flow, such as at a pipe junction, Eq. 1.14 can be generalized to

$$\Sigma \rho_i A_i V_i = 0 \tag{1.15}$$

which when applied to pipes means that the summation (or integration) of the mass flux through the pipes at a junction must be zero. This approach works well for any situation where areas are easily calculated and the velocity is uniform and normal to the area.

The continuity equation for steady incompressible fluid can be made more general by considering a control surface $cs$ that completely encompasses the region of interest. The surface integral of the mass flux through the control surfaces must be zero. Since the local velocity may not be normal to its corresponding incremental area, the vector dot product of the velocity vector $V$ and the element of area $dA$ (a vector quantity) must be used. The continuity equation becomes

$$\int_{cs} \rho V \cdot dA = 0 \tag{1.16}$$

The restriction on Eq. 1.16 is that there is no change in storage within the control volume $cv$ which is bounded by the control surface $cs$. The final step in completely generalizing the continuity equation is to remove this limitation. Instead of being equal to zero, the net mass flux through the control surface must exactly equal the time rate of change of mass inside the control volume or

$$\int_{cs} \rho V \cdot dA = \frac{\partial}{\partial t} \int_{cv} \rho \, dv \tag{1.17}$$

in which $v$ is volume.

The preceding five forms of the continuity equation are in an integral form and apply to finite areas or volume. If information is desired on a very small elemental area or a small control volume, the differential form is necessary. This form of the equation, in Cartesian coordinates, is developed by considering a cubical control volume whose dimensions are $\Delta x$, $\Delta y$, $\Delta z$. The net mass flux in each coordinate direction through the appropriate elemental flow surfaces is calculated. The change of storage or compressibility must also be considered. The result is

$$\frac{\partial \rho}{\partial t} + \frac{\partial(\rho u)}{\partial x} + \frac{\partial(\rho v)}{\partial y} + \frac{\partial(\rho w)}{\partial z} = 0 \tag{1.18}$$

in which $u$, $v$ and $w$ are the components of velocity in the $x$, $y$ and $z$ coordinate directions. If the flow is steady, the first term becomes zero. If the flow is steady and the fluid incompressible, $\rho$ can be eliminated and the equation simplified to

$$\partial u/\partial x + \partial v/\partial y + \partial w/\partial z = 0 \tag{1.19}$$

## Momentum Equation

This equation, sometimes referred to as the equation of motion or the linear momentum equation, is derived from Newton's second law. The principle can be stated as follows: The net force acting on a control volume is equal to the net momentum flux through the control surface plus the time rate of change of momentum inside.

Since momentum is the product of mass and velocity, it is a vector quantity acting in the same direction as the velocity vector. To apply it, one must first identify a control volume. Consider steady two-dimensional flow in a pipe elbow. The control volume is made up of the physical boundaries of the elbow and two cross sectional areas 1 and 2 cut through the elbow such that they are normal to the local velocity vectors $V_1$ and $V_2$. The velocity profile is assumed to be uniform at sections 1 and 2. Since the flow is steady, the time rate of change of momentum is zero. The momentum flux at the two stations is the vector dot product of the mass flow and the local velocity vector. The $x$ component of the force required to hold the elbow in equilibrium plus the $x$ component of the pressure force at the two cut sections is equal to the net momentum flux in the $x$ direction.

$$\Sigma F_x = \rho_2 A_2 V_2 V_{2x} - \rho_1 A_1 V_1 V_{1x} \tag{1.20}$$

For incompressible flow, this equation can be reduced to

$$\Sigma F_x = \rho Q(V_{2x} - V_{1x}) \tag{1.21}$$

These equations can easily be applied to a three dimensional flow problem by adding equations in the $y$ and $z$ directions.

If the flow is unsteady or if it is not possible to select the control volume so that the velocity vector is everywhere normal to the control surface, the general integral form of the momentum equation should be used. The $x$ component is

$$\Sigma F_x = \partial/\partial t \int_{cv} \rho u \, dv + \int_{cs} \rho u \mathbf{V} \cdot d\mathbf{A} \tag{1.22}$$

The other two components are similar. For application to flow in rigid conduits, it is usually not necessary to formally integrate Eq. 1.22. An example of this is in Chapter 8, where the unsteady momentum equation is used to develop the basic equations for analyzing hydraulic transients.

**Energy equation**

The first law of thermodynamics states that the change of internal energy of a system is equal to the sum of the energy added to the fluid and the work done by the fluid. A general form of the energy equation for incompressible pipe flow (assuming a uniform velocity profile) is

$$\frac{P_1}{\rho} + gz_1 + \frac{V_1^2}{2} = \frac{P_2}{\rho} + gz_2 + \frac{V_2^2}{2} - W_p + W_t + W_f \tag{1.23}$$

The units of each term are energy per unit mass. The first two terms on both sides of the equation are potential energy, the third term is kinetic energy, $W_p$ is pump energy added to the system, $W_t$ is turbine energy removed from the system, and $W_f$ represents friction and other minor energy losses. Equation 1.23 is restricted to steady flow and ignores nuclear, electrical, magnetic and surface-tension energy.

An alternate form of the energy equation is obtained by dividing Eq. 1.23 by gravity. The units are energy per unit weight of liquid: ft-lb/lb or N-m/N, which reduces to ft or m, respectively, after simplification. This form of the equation is

$$\frac{P_1}{\gamma} + z_1 + \frac{V_1^2}{2g} = P_2/\gamma + z_2 + \frac{V_2^2}{2g} - H_p + H_t + H_f \tag{1.24}$$

The first three terms are pressure head ($P/\gamma$), elevation head ($z$) above some datum, and velocity head ($V^2/2g$). The last three terms on the right side of the equation are the total dynamic head added by a pump ($H_p$) or removed by a turbine ($H_t$) and the friction plus minor head losses ($H_f$). If there is no pump or turbine and if friction losses are negligible, the last three terms disappear and Eq. 1.24 reduces to the Bernoulli equation.

Energy is a scalar quantity and the energy equation is a point function. This means that when comparing the energy at any two points in a system, any path taken to get from point 1 to 2 must produce the same energy at point 2. This is a basic principle used in pipe network analysis.

When applying the energy equation to pipelines, it is convenient and instructive to show it graphically. This is demonstrated with a simple system shown in Fig. 1.2. Six locations have been identified. At each point, the various terms of the energy equation are shown by arrows above the reference datum. At point 1, on the surface of the upstream reservoir, the gauge pressure is zero (atmospheric) and there is no velocity. The entire energy is $z_1$. At point 2, also in the reservoir but at the elevation of the pipe outlet, the total energy is $z_2 + P_2/\gamma$. At point 3, just inside the pipe, the fluid has attained a velocity and hence some kinetic energy. The total energy is $z_3 + P_3/\gamma + V_3^2/2g$. Ignoring minor losses, the total energy is the same at sections 1, 2, and 3. One point to note is how the division between the three types of energy depends on the location.

As the water moves through the pipe, the total energy reduces due to friction. The energy must eventually equal $z_6$, which is the total energy in the second reservoir. If the pipe has a constant diameter, the velocity will be the same at all sections and hence the velocity head $(V^2/2g)$ will be equal at sections 3, 4, and 5. If the pipe roughness is uniform, then each length of

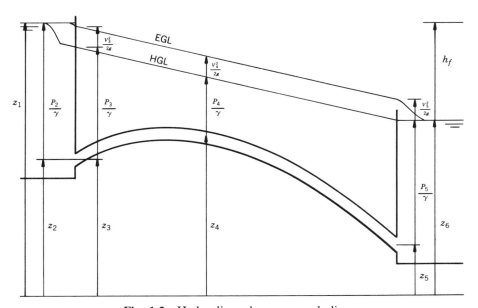

**Fig. 1.2** Hydraulic and energy grade lines.

pipe will have the same friction loss. The system will self adjust until the friction and minor losses just equal $z_1 - z_6$.

The line connecting points of total energy is called the *energy grade line* (EGL). For the case just described, the EGL has a constant slope. The line connecting the points of elevation plus pressure is called the *hydraulic grade line* (HGL). For the example in Fig. 1.2, the lines are parallel because the pipe has a constant diameter and roughness.

The height of the EGL above the datum is commonly referred to as the *total head*, and the height of the HGL above the datum is called the *piezometric head*.

A simple application of the energy equation is a flow measuring venturi, as shown in Fig. 1.3. A venturi is a contracted section of pipe designed to prevent flow separation and minimize hydraulic losses. The constriction creates a locally high velocity and a correspondingly reduced pressure. Properly designed, there is negligible loss between sections 1 and 2 so that

$$\frac{P_1}{\gamma} + z_1 + \frac{V_1^2}{2g} = \frac{P_2}{\gamma} + z_2 + \frac{V_2^2}{2g}$$

The form of the equation is the same, regardless of the datum or the orientation of the venturi. Expressing $V_1$ and $V_2$ as a function of $Q$ and areas $A_1$ and $A_2$,

$$\Delta H = (P_1/\gamma + z_1) - (P_2/\gamma + z_2) = \frac{Q^2}{2g}(1/A_2^2 - 1/A_1^2)$$

The left side of the equation is the difference in the HGL (or piezometric heads) between points 1 and 2 in Fig. 1.3. This head change $\Delta H$ is easily measured by a differential manometer and is independent of the orientation

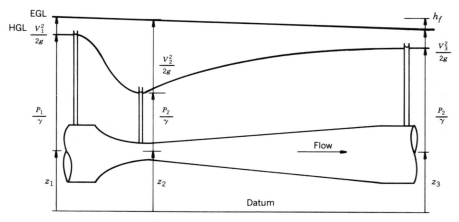

**Fig. 1.3**   Venturi flow meter.

of the venturi. The equation for flow through a venturi the preceeding equation reduces to

$$Q = C\sqrt{\Delta H}\ .$$

## 1.4   FLUID FRICTION

Application of the energy equation requires an accurate estimate of the energy losses caused by shear stress between the fluid and the boundary. Eq. 1.12 identifies that the shear stress is a function of viscosity and the velocity gradient near the boundary. The velocity gradient is controlled by the velocity, the boundary roughness, and thickness of the boundary layer.

Consider turbulent flow in the entry region of a rough pipe. If flow is provided from a large chamber with a well designed nozzle, the velocity distribution at the entrance to the pipe will be uniform except for a thin boundary layer. Near the boundary, the velocity gradient and consequently the shear stress will be maximal. With increasing distance, the boundary layer thickens due to momentum interchange caused by turbulent eddies generated in the boundary layer. This reduces the velocity gradient and local shear stress. This continues until the thickness of the boundary layer equals the radius of the pipe. Beyond that point, only minor changes occur in the velocity profile as it evolves into its fully developed shape.

The variation of relative wall shear stress $\tau/\tau_0$ in a pipe has shown that near the pipe entrance the local shear $\tau$ is about 2.5 times the fully developed value $\tau_0$ (78). It reduces rapidly with distance, approaching $\tau_0$ in about 15 pipe diameters. The velocity profile takes longer to develop. If the pipe entrance is not well rounded and flow separation occurs (as for an abrupt or reentrant condition), the velocity profile and wall shear stress reach their steady-state values in a shorter length. This is also the case for flow downstream from a valve or orifice. The high level of local turbulence is efficient in reestablishing the velocity profile and wall shear stress.

Another case of nonuniform flow where local shear stress can be affected is downstream from two elbows out of plane, or a valve and elbow combination. For such cases, secondary flows are established that can produce a significant velocity component in the circumferential direction. This increases the velocity near the wall and increases the shear stress. This spiral flow can persist for many pipe diameters.

Although of fundamental interest, nonuniform flow and its influence on wall shear stress is seldom a significant problem in pipeline design. One reason is that most systems are long enough that the length influenced by a variable $\tau$ is insignificant. If the system is short, it frequently contains many bends, valves, transitions, etc., whose influence completely overshadows the friction losses. The most significant problem with pipeline design is to obtain a reliable value of the shear stress or pipe friction factor for fully developed flow or the loss coefficients for local losses.

From an engineering point of view, it is not practical to work in terms of wall shear stress since it requires detailed information on the velocity gradient. The velocity gradient not only varies with distance in developing flow, but it is also a function of velocity and viscosity for fully developed flow. It is easier to work in terms of the average shear stress or friction loss over a length of pipe. The friction loss between two points in a pipe is equal to the decrease in the total head or EGL. Dimensional analysis can be used to provide a functional relationship between the friction loss $h_f$, the important fluid properties, and flow parameters. The resulting equation is called the Darcy–Weisbach equation:

$$h_f = \frac{fLV^2}{2gd} = \frac{fLQ^2}{1.23gd^5} \tag{1.25}$$

The friction factor $f$ has been evaluated experimentally for numerous pipes. Such tests have shown $f$ to be a function of pipe diameter, roughness, and Reynolds number Re. Since roughness may vary with time due to buildup of solid deposits or organic growths, $f$ is also time dependent. Manufacturing tolerances also cause variations in the pipe diameter and surface roughness. The point that is being made is that it is really not possible to know the friction factor of any pipe precisely. A designer is required to use good engineering judgment in selecting a design value for $f$ so that proper allowance is made for these factors.

The functional relationship of $f$ with roughness $e$, diameter $d$, and Re has been investigated quite thoroughly. The pioneering work was done by Nikuradse (41) and Colebrook (12). Their work is the basis of the Moody chart (38), Fig. 1.4, which provides a graphical means of evaluating $f$. The chart includes: 1) the laminar region where $f = 64/Re$; 2) the critical region, $2000 < Re < 4000$, where a unique value of $f$ does not exist; 3) the transition region between a smooth pipe and a fully rough pipe where $f$ is a function of $e/d$ and Re; and 4) the fully turbulent zone where $f$ is a function of $e/d$ only.

Using the Moody diagram to get $f$ requires that Re and $e/d$ be known. Calculating Re is easy if the temperature, velocity, and pipe diameter are known. The problem is obtaining a good value for $e$, the roughness height. Typical values of $e$ are listed in the lower left-hand corner of Fig. 1.4. For other values, see page 6.9 of Reference 31. These values should be considered as guides only and not be used if more exact information is available. The value of $e$ for commercial pipe is not directly measurable, so it is evaluated indirectly. The procedure is to test several lengths of a given pipe over a range of Reynolds numbers to get $f$ versus Re. By plotting the data on the Moody diagram, the value of $e/d$ can be determined.

The original Colebrook equation used to develop the Moody diagram was in implicit form and therefore required an iterative solution. Wood (85)

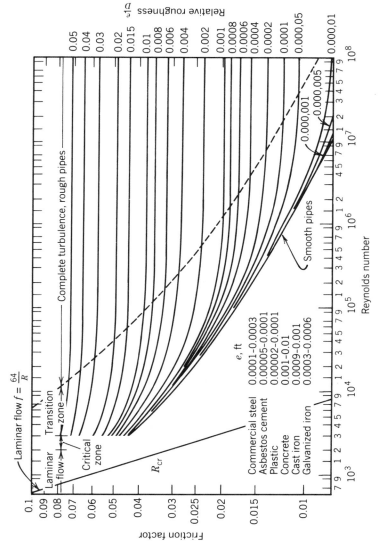

**Fig. 1.4** The Moody diagram.

developed an explicit form of the Colebrook equation which closely approximates it for $\mathrm{Re} > 10^4$ and $10^{-5} < k < 0.04$ ($k = e/d$):

$$f = a + b\,\mathrm{Re}^{-c}$$

$$a = 0.094k^{0.225} + 0.53k, \; b = 88k^{0.44}, \; c = 1.62k^{0.134}$$

(1.26)

Such an equation is especially helpful for computer solutions, since it actually stores most of the Moody chart in an easily retrievable form. For large-diameter pipes, Reference 76 provides results of numerous field tests of $f$. This reference is helpful in providing more confidence in estimating $f$ for large pipes.

A relationship between $f$ and $\tau$ can be developed by applying the momentum equation to steady flow in a pipe. Consider the short pipe segment shown in Fig. 1.5 and apply the steady momentum equation with the direction of flow as the $x$ direction.

$$P_1 A - P_2 A - \tau_0 \pi\, dL - W \sin \alpha = 0$$

The weight component can be simplified by

$$W \sin \alpha = \gamma A L \sin \alpha = \gamma A (z_2 - z_1)$$

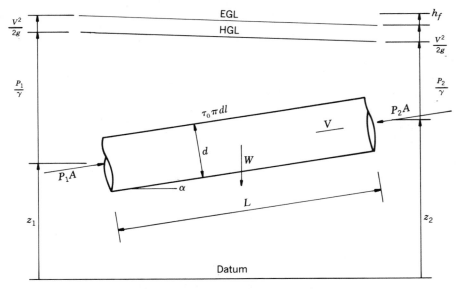

**Fig. 1.5**   Steady-state flow in a pipe.

Combining this with the pressure terms, dividing by $\gamma A$ and rearranging,

$$\left(\frac{P_1}{\gamma} + z_1\right) - \left(\frac{P_2}{\gamma} + z_2\right) = 4\tau_0 L/\gamma d$$

The left side is the difference in piezometric head which is defined as the friction loss $h_f$, or the drop in the EGL as shown in Fig. 1.2. This loss is the same $h_f$ calculated from the Darcy-Weisbach equation (Eq. 1.25). Substituting in Eq. 1.25 and rearranging provides the following functional relationship:

$$f = 8\tau_0/\rho V^2 \tag{1.27}$$

**Application of the Darcy Equation**

Problems involving pipe friction generally fall into one of three categories. The steps involved in solving these three cases are summarized. It is assumed that in each case the water temperature and, hence, the viscosity is known.

CASE 1. *Given*: $L$, $d$, $e$, $T$, and $V$; *Find*: $h_f$. This case is simple because it has a direct solution involving only two steps:

  (a) Calculate Re and $e/d$ and find $f$ from the Moody chart (Fig. 1.4) or from Eq. 1.26.
  (b) Using Eq. 1.25, calculate $h_f$.

CASE 2. *Given*: $L$, $d$, $e$, $T$ and $h_f$; *Find*: $V$.   This case requires a trial and error solution because $f$ is a function of an unknown $V$. There are a number of ways to solve it. The following is but one. The solution can be programmed into a hand-held calculator, computer, or done with a spreadsheet using a microcomputer.

  (a) Assume a reasonable velocity (5–10 fps,) and calculate Re.
  (b) Calculate $e/d$ and find $f$ from the Moody chart or Eq. 1.26.
  (c) Using Eq. 1.25, calculate the velocity corresponding to the available $h_f$.
  (d) Compare the calculated $V$ with the assumed value. If it is significantly different calculate a new Re, pick a new $f$ and repeat the process.

CASE 3. *Given*: $L$, $e$, $Q$, $T$, and $h_f$; *Find*: $d$.   This case is typical of the first step in designing a pipeline where the available head drop and flow rates are known and the pipe diameter is needed. This solution is also trial and error because $f$ is a function of $V$, which is unknown. Again, several solution schemes can be used. One is suggested. If $f$ is known or assumed, then the solution is direct. Remember that only commercially available pipe diameters can be used.

(a) Using Eq. 1.25 and a trial value of $f = 0.01$ to $0.02$, calculate a trial diameter $d$ and select the next largest available pipe diameter.

(b) Calculate $e/d$, $V$, and Re and get an adjusted $f$ from Fig. 1.4 or Eq. 1.26.

(c) Check $h_f$ using Eq. 1.25. If the calculated $h_f$ is less than the available $h_f$, the pipe is adequate. If there is a large difference in the two $h_f$ values, check the next smallest pipe diameter.

(d) Select the smallest pipe that produces a $h_f$ less than or equal to the available $h_f$. It is seldom possible to exactly match the calculated and available $h_f$ values. To make them match would require a non-standard pipe size.

In all of the above examples, it is necessary to allow for aging and uncertainties involved in determining the friction factor. This can be done by increasing the $f$ factor obtained from the Moody chart or allowing an adequate difference between the calculated and available head losses. If it is necessary to have the exact discharge, the difference can be made up by a control valve so that the required flow rate is provided. As the pipe ages, less pressure drop would be required by the valve.

**Other Empirical Formulas**

The Darcy–Weisbach equation (Eq. 1.25) is commonly used in academic circles to calculate head loss because it is the most exact. This is because the variation of $f$ with pipe roughness and Reynolds number is properly accounted for when the Moody diagram (Fig. 1.4) or Woods' equation (1.26) is used. Two other equations in use are the Hazen–Williams and Manning formulas.

<u>Hazen–Williams</u>

$$\text{English} \qquad V = 1.318 C_h \, R^{0.63} S^{0.54} \qquad (1.28a)$$

$$\text{Metric} \qquad V = 0.85 C_h R^{0.63} S^{0.54} \qquad (1.28b)$$

<u>Manning</u>

$$\text{English} \qquad V = \frac{1.486}{n} R^{0.667} S^{0.5} \qquad (1.29a)$$

$$\text{Metric} \qquad V = \frac{1}{n} R^{0.667} S^{0.5} \qquad (1.29b)$$

In the above equations, $R$ is the hydraulic radius = area/wetted perimeter ($R = d/4$ for circular pipe flowing full), $S$ is the slope of the energy grade line = $h_f/L$, and $C_h$ and $n$ are roughness factors or friction coefficients.

The friction coefficients $C_h$ and $n$ are assumed to be independent of Reynolds number. Even though this is not correct, the uncertainty involved in obtaining a reliable value of the friction coefficients can introduce an error as large as the variation with Reynolds number.

Engineers in the waterworks field generally use the Manning or the Hazen–Williams equation. This choice may have been based on simplifications of their solution due to the availability of special slide rules, tables, nomographs, and charts. The availability of modern hand calculators and computers makes the solution of any of the equations so simple that this is no longer a factor. Even so, it is unlikely that there will be any significant change in the preferred equation by the various user groups. It is therefore advisable to be familiar with all three equations.

Manning's $n$ values are listed in Table 1.2. Extreme values of $n$ range from 0.01 for new pipe in excellent condition to 0.035 for an extremely rough pipe. Typical values used in design range from 0.011 to 0.017. The same caution should be exercised ʾ      ʾhese values as was suggested for the tabulated $f$ values in Fig. 1.4. W            ⁓ssible, obtain information from the manufacturer. Suggested            -Williams $C_h$ are listed in Table 1.3. Values of $C_h$ rangⁿ            ⁓ in excellent condition to less than 100 for old piʾ            ⁏al values would be between 120 and 130.

**TABLE 1.2   Values of**

| Kind of Pipe | .iation | | Use in Designing | |
| --- | --- | --- | --- | --- |
| | ɑ | To | From | To |
| Clean uncoated cast-iron pipe | .011 | 0.015 | 0.013 | 0.015 |
| Clean coated cast-iron pipe | ɔ.010 | 0.014 | 0.012 | 0.014 |
| Dirty or tuberculated cast-iron pipe | 0.015 | 0.035 | | |
| Riveted steel pipe | 0.013 | 0.017 | 0.015 | 0.017 |
| Lock-bar and welded pipe | 0.010 | 0.013 | 0.012 | 0.013 |
| Galvanized-iron pipe | 0.012 | 0.017 | 0.015 | 0.017 |
| Brass and glass pipe | 0.009 | 0.013 | | |
| Wood-stave pipe | 0.010 | 0.014 | | |
| Wood-stave pipe, small diameter | | | 0.011 | 0.012 |
| Wood-stave pipe, large diameter | | | 0.012 | 0.013 |
| Concrete pipe | 0.010 | 0.017 | | |
| Concrete pipe with rough joints | | | 0.016 | 0.017 |
| Concrete pipe, "dry mix," rough forms | | | 0.015 | 0.016 |
| Concrete pipe, "wet mix," steel forms | | | 0.012 | 0.014 |
| Concrete pipe, very smooth | | | 0.011 | 0.012 |
| Virtrified sewer pipe | 0.010 | 0.017 | 0.013 | 0.015 |
| Common clay drainage tile | 0.011 | 0.017 | 0.012 | 0.014 |

*Source*:   R. Manning, "Flow of Water Open Channels and Pipes," *Trans. Inst. Civil Engrs.* (*Ireland*), vol. 20, 1890.

**TABLE 1.3   Values of Hazen–Williams coefficient $C_h$** [a]

| Character of Pipe | Hazen–Williams Coefficients of Roughness $C_h$ |
|---|---|
| New or in excellent condition cast-iron and steel pipe with cement, or bituminous linings centrifugally spun, cement–asbestos pipe, copper tubing, brass pipe, plastic pipe, and glass pipe | 140 |
| Older pipe listed above in good condition, and/or pipes, cement mortar lined in place with good workmanship, larger than 24 in. in diameter | 130 |
| Cement mortar lined pipe in place, small diameter with good workmanship or large diameter with ordinary workmanship: work stave; tar-dipped cast-iron pipe new and/or old in inactive water | 120 |
| Old unlined or tar-dipped cast-iron pipe in good condition | 100 |
| Old cast-iron pipe severely tuberculated or any pipe with heavy deposits | 40–80 |

[a]Reference 3. (Courtesy the American Society of Civil Engineers.)

### Limiting Velocities

For systems where the pipes are long and the available head is limited, the calculations may produce a velocity that is too low. If the velocity is below about 3 fps, problems may develop due to suspended solids settling out or trapped air that cannot be removed. The safe lower limit to avoid sedimentation depends on the amount and type of sediment. The velocity required to remove air is generally about 2–3 fps. It varies some with the pipe diameter and slope. The velocity to remove air is generally larger than the sedimentation velocity. Usually, there is no problem with minimum velocity at the design flow rate because normal velocities are in the range of 5–10 fps. The problem arises when the pipe is operated at low flow rates. If long periods of low flows are anticipated some provision should be made to flush at higher velocities periodically.

It is also necessary to consider the safe upper limit of the velocity. Problems associated with high velocities are: 1) erosion of the pipe wall or liner (especially if coarse suspended sediment is present), 2) cavitation, 3) increased pumping costs, 4) hydraulic transients. Criteria for cavitation and transient analysis are discussed in later chapters.

For short high head gravity flow systems, when the pipe diameter is increased to limit the maximum velocity, some type of energy dissipator (valve, orifice, etc.) must be installed to make up for the decreased friction loss.

## 1.5  MINOR LOSSES

Minor losses is a term referring to losses that occur at a pipe entrance, elbow, orifice, valve, etc. They are more appropriately called local losses. The term minor loss is really only appropriate if the summation of all the local losses produce a head loss that is small compared with the friction loss.

For short systems, friction may be negligibly small compared with the minor losses. Judgment must be used in deciding how important the minor losses are and, therefore, how much effort should be expended in evaluating the various loss coefficients.

Energy dissipation, or head loss, caused by a local disturbance is due primarily to the formation and decay of turbulent eddies. Consider flow through an orifice as an example. Separation of flow at the lip of the orifice plate causes an intense shear layer between the jet and the separation zone. This creates eddies having high rotational velocities. These eddies are destroyed by viscous shear and interaction with other eddies, causing a loss of energy. This is discussed in more detail in Chapter 5.

The head loss $h_l$ caused by a minor loss is proportional to the velocity head.

$$h_l = \frac{K_l Q^2}{2gA^2} \tag{1.30}$$

The loss coefficient $K_l$ is analogous to $fL/d$ in Eq. 1.25. In fact, some prefer to express loss coefficients as an equivalent pipe length: $L/d = K_l/f$. It simply represents the length of pipe that produces the same head loss as the minor loss. This is a convenient means of including minor losses in the Hazen–Williams and Manning equations. The friction slope $S$ in Eqs. 1.28 and 1.29 is calculated by dividing the available head drop by the sum of the actual pipe length and the equivalent length of all minor losses. Nomographs for finding the equivalent length for several fittings are available (3 and 13). It is important to note that there is not one unique equivalent length of pipe for a given minor loss, since it depends on the pipe roughness. When these losses are truly minor, this problem becomes academic because the error only influences losses which make up a small percentage of the total. For cases where accurate evaluation of all losses is important, it is recommended that the Darcy equation and the minor loss coefficients be used.

For use with the Darcy equation, $K_l$ is used rather than equivalent length. The total head loss term in the energy equation can be written as

$$H_f = \left( \Sigma \frac{fL}{2gdA^2} + \Sigma \frac{K_l}{2gA^2} \right) Q^2 = CQ^2 \tag{1.31}$$

The summation term represents the numerical sum of all minor loss coefficients. If the minor loss is different in diameter than the pipe, the proper area in Eq. 1.31 must be used.

Typical values of loss coefficients for various minor losses are summarized in Table 1.4. For tees and diffusions, there are too many variables to give a single value. See Reference 37 for details.

**TABLE 1.4  Minor Loss Coefficients**

| | $K_l$ | |
|---|---|---|
| Item | Typical Value | Typical Range |
| Pipe inlets | | |
|   Inward projecting pipe | 0.78 | 0.5 to 0.9 |
|   Sharp corner—flush | 0.50 | |
|   Slightly rounded | 0.20 | 0.04 to 0.5 |
|   Bell mouth | 0.04 | 0.03 to 0.1 |
| Expansions[a] | $(1 - A_1/A_2)^2$ (based on $V_1$) | |
| Contractions[b] | $(1/C_c - 1)^2$ (based on $V_2$) | |

| $A_2/A_1$ | 0.1 | 0.2 | 0.3 | 0.4 | 0.5 | 0.6 | 0.7 | 0.8 | 0.9 |
|---|---|---|---|---|---|---|---|---|---|
| $C_c$ | 0.624 | 0.632 | 0.643 | 0.659 | 0.681 | 0.712 | 0.755 | 0.813 | 0.892 |

| Item | Typical Value | Typical Range |
|---|---|---|
| Bends[c] | | |
|   Short radius, $r/d = 1$ | | |
|     90 | | 0.24 |
|     45 | | 0.1 |
|     30 | | 0.06 |
|   Long radius, $r/d = 1.5$ | | |
|     90 | | 0.19 |
|     45 | | 0.09 |
|     30 | | 0.06 |
|   Mitered (one miter) | | |
|     90 | 1.1 | |
|     60 | 0.55 | 0.4  to 0.59 |
|     45 | 0.4 | 0.35 to 0.44 |
|     30 | 0.15 | 0.11 to 0.19 |
| Tees | [c] | |
| Diffusers | [c] | |
| Valves | | |
|   Check valve | 0.8 | 0.5 to 1.5 |
|   Full open gate | 0.15 | 0.1 to 0.3 |
|   Full open butterfly | 0.2 | 0.2 to 0.6 |
|   Full open globe | 4.0 | 3 to 10 |

[a]Reference 58, p. 170.
[b]Reference 58, p. 305.
[c]Reference 37.

One other factor that is important if minor losses become significant is the interaction between components placed close together (see Reference 37). Depending on the type and spacing of the components, the total loss coefficient may be greater or less than the simple sum of the individual values.

## 1.6   INTRODUCTION TO CAVITATION

Cavitation is discussed in Chapters 5–7. Since it will come up frequently in the material presented in Chapters 2–4, it is appropriate to describe it briefly in this chapter. Cavitation is caused by rapid vaporization and condensation of a liquid. It originates from voids or tiny bubbles containing gas and vapor which form a nucleus for vaporization. If these bubbles are subjected to vapor pressure, they grow rapidly. If the surrounding pressure is above vapor pressure, they become unstable and collapse violently.

The low pressures necessary for cavitation to form and the high pressures causing collapse are frequently associated with the formation and decay of turbulent eddies in boundary layers or separation regions. These are the same eddies identified as the primary cause of energy dissipation at local disturbances (or minor losses). The low pressure at the center of these eddies, caused by their high rotational velocity, combined with the generally low pressure of the separation region, can cause cavitation. The collapse of the cavities causes noise, pressure fluctuations, vibrations, and possible erosion damage. When cavitation becomes severe, the large amount of vapor can cause erosion damage or reduce the efficiency of hydraulic equipment and machinery.

In succeeding chapters, cavitation will be included as an important design parameter affecting selection and operation of pumps, valves, and other control devices. It is one of the hydraulic problems that has caused costly repairs, replacement, and in some cases catastrophic failures of hydraulic structures.

## 1.7   INTRODUCTION TO HYDRAULIC TRANSIENTS

Like cavitation, hydraulic transients will be discussed in subsequent chapters. It is a complex topic and requires several chapters to cover it even partially. Since it is also mentioned frequently as a serious hydraulic problem in Chapters 2–4, it will be briefly introduced.

Hydraulic transients, in the most general term, refer to any unsteady flow in open channel or closed conduits. It can be subdivided, according to the rate of change of velocity and the mathematical means of analysis, into surges and transients (or waterhammers). During a surge, the velocity changes slowly with time and the entire body of fluid can be considered as moving as a solid body. A simple example is an oscillating U-tube.

For a transient or waterhammer in a closed conduit, the velocity changes rapidly. An example is instant (or rapid) closure of a valve. When a valve is rapidly closed, the force needed to destroy the original momentum of the flowing liquid causes a high pressure. This pressure is transmitted through the conduit at the acoustic wave speed and can rapidly subject the entire pipe to an increased pressure.

The magnitude of the pressure rise is dependent on the velocity change and the acoustic wave speed. For a typical steel pipe, a decrease of 1 fps (0.305 m/s) in the velocity causes the head to increase about 100 ft (30.5 m) or 23.1 psi (160 kPa). Since normal pipe velocities are in the 5–10 fps (1.5–3 m/s) range, rapidly stopping the flow can easily cause excessive pressures.

Typical causes of transients include filling and flushing pipes, opening or closing valves, starting or stopping pumps, and air moving through pipes or being released incorrectly. Transients constitute a subject that is generally inadequately covered in design specifications and yet they are responsible for many pipe and equipment failures.

## PROBLEMS FOR CHAPTER 1

**1.1.** Water at 70°F is flowing in a 24-in. diameter pipe at 1.5 fps. (a) Calculate Re. (b) Find the velocity above which turbulent flow exists.

*Answer:*   $Re = 2.83 \cdot 10^5$, $V = 0.021$ fps

**1.2.** List three examples of: (a) gradually varied pipe flow, (b) rapidly varied pipe flow, (c) surges, and (d) transients.

**1.3.** A test is made with water to measure its bulk modulus $K$. The test results showed that a pressure increase of 1000 psi reduced the volume by 0.7%. What is the value of $K$? What might explain its low value?.

*Answer:*   $K = 143,000$ psi

**1.4.** Calculate the acoustic wave speed for water in a rigid pipe at $T = 100°F$. Calculate in English and SI units.

*Answer:*   $a = 4,943$ fps $= 1,508$ m/s

**1.5.** A valve is operating in a system where $P_u = 75$ psi and $\Delta P = 50$ psi. The water temperature is 40°F and the elevation of the installation is 4,000 ft. Calculate $\sigma$ (eq. 1.10b) using $P_u$ as the reference pressure.

*Answer:*   $\sigma = 1.75$

**1.6.** Using the energy and momentum equations derive equations to calculate $P_2$ and $P_3$ for the nozzle sudden expansion shown below. $P_3$ is beyond the zone of flow establishment.

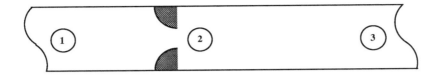

**1.7.** Draw the hydraulic and energy grade lines for the system on the handout. Show all minor losses.

**1.8.** Compare $f$ found from Fig. 1.4 with $f$ calculated from Eq. 1.26 for a pipe with $e = 0.003$ ft, $D = 12$ in., and $T = 60°F$. Check for $V = 0.3$ fps, and 10 fps.

*Answer*:   $f = 0.0307, 0.0307, f = 0.0273, 0.0265$

**1.9.** Calculate the wall shear stress for problem 1.8 at $V = 10$ fps. Calculate in English and SI units at 60°F, 15.5°C.

*Answer*:   $\tau = 0.642$ lb/sq ft $= 30.75$ Pa

**1.10.** Water at 20°C is carried by gravity flow between two reservoirs through 1000 m of 305-mm diameter pipe in series with 500 m of 203-mm diameter pipe. Assume an average roughness value for commercial steel pipe. The minor losses include a square edge entrance, 4–90 short radius elbows, the sudden contraction from 305 to 203 mm, one globe valve, and the pipe exit. The difference in elevation between the two reservoirs is 150 m. Calculate the discharge. The minor losses are all in the 203 mm pipe (except the entrance).

*Answer*:   $Q = 0.232$ m$^3$/s

**1.11.** Using the following data for a pipe, calculate $f$ and $C_h$ at $V = 2$ and 20 fps. Does $C_h$ vary with Re, $\nu = 0.00001$ ft$^2$/s, $e/D = 0.0001$, $D = 9$ in.

*Answer*:   at $V = 2$, $f = 0.0173$, $C_h = 150$

at $V = 20$, $f = 0.013$, $C_h = 145$,

**1.12.** Write a program in FORTRAN or for a spreadsheet to iterate to evaluate the pipe diameter given: $L$, $e$, $Q$ and $h_f$.

# 2

# DESIGN OF PIPELINES

Designing water conveyance systems is a complex process involving much more than a simple determination of the required pipe size. One must consider social, environmental, political, and legal problems as well as the engineering design. The time is past when projects could be built with little or no regard to their impact on the social well-being of the public and on the environment. Protecting the quality of our life and environment is and will continue to be a major factor in determining the fate of future projects. This chapter first considers the feasibility and economics of building a pipeline. Next, pipe materials and pressure rating are discussed, followed by examples of selecting the correct pipe diameter. Factors such as aging of the pipe and maintaining positive pressure are included. Examples demonstrate selection of pipe diameter for gravity flow and for pumped systems where pipe cost must be balanced with pumping costs.

## 2.1 FEASIBILITY STUDY

The first step in any design is to conduct a feasibility study. Such studies have no set format or a firm list of items to include. In general, it consists of making a preliminary design to identify the project scope and all major features that would influence the cost or viability. The proposed design is then analyzed relative to legal, political, social, environmental, and economic aspects.

A number of items that should be considered in a feasibility study are listed in Table 2.1. So many site-specific or time-dependent factors exist that

**TABLE 2.1  Feasibility Checklist**

A. Preliminary design
 1. Identify service area and water demands
 2. Availability of adequate water supply
 3. Requirements for water treatment
 4. Identify major pipe routes and any special geologic features or major crossings
 5. Determine number, location, and size of water storage facilities
 6. Estimate pumping requirements and number of control structures
 7. Anticipate special problems, such as transients, cavitation, and winterizing, if they have a significant impact on the cost of the project

B. Legal problems
 1. water rights
 2. right-of-ways
 3. safety
 4. reliability

C. Social problems
 1. General public reaction to project
 2. Impact on local economy, either long-term or short-term
 3. Impact on quality of life, especially if it is a major project involving a large transient work force
 4. Noise and vibration generated at pump stations or control structures
 5. Location and general appearance of any structures such as water tanks, reservoirs, pumping plants, and control structures
 6. Safety-related problems
 7. Impact on local water supplies

D. Environmental concerns
 1. Physical damage to the terrain, including visual damage, potential erosion, and damage to vegetation; this would include the pipeline routes, access roads, storage areas, reservoirs, etc.
 2. Impact on wildlife
 3. Water or air pollution

E. Economic analysis

such a list can only serve as a starting point. Local laws, social values, and environmental concerns can vary significantly between geographic areas. Engineering problems in the mountainous western areas pose totally different design constraints than would those in the great plains area. An engineer must be aware of the problems unique to the area.

If adequate attention is given to the social and economic impact of a project at the earliest possible stage, it may be possible to avoid misunderstandings and costly delays caused by public opposition. It is the responsibility of all to protect and improve the quality of our life, environment, and natural resources. At the same time, we must accommodate the increasing demand for energy and water required for continued economic growth. Improved technology and greater sensitivity to public needs and desires are needed to accomplish these objectives.


## 2.2  ECONOMIC ANALYSIS

In any pipeline design there are numerous choices available which affect the economics of the project. Typical examples include alternative pipe routes, amount of storage and its effect on reliability and controllability of flow, pipe material and diameter, provision for future demands, etc. In making decisions, both the engineering and economic advantages of the alternatives must be considered. In most cases, the final design is based on selecting the most economical solution, as long as it is technically sound. The design that is most economical may not be the cheapest or the best hydraulically. Reliability, safety, maintenance, operating, and replacement costs must all be given their proper value. The final decisions should be based on input from qualified individuals knowledgeable about these factors.

An accurate economical analysis of a complex system which includes all important variables can only be done with "systems approach." This approach involves identifying all variables and alternatives that have an impact on the cost and developing functional relationships and solving them simultaneously with any of a number of computer techniques to optimize the design and minimize cost. This is a complex process and a complete discussion is beyond the scope of this text. Those desiring more detail should consult texts on the subject (References 27 and 54).

The intent of this section is primarily to discuss the principles rather than the procedures of an economic analysis. These principles will serve as background data for developing the objective functions required for the systems analysis. A few numerical examples are presented to demonstrate the principles. Because costs and interest rates change rapidly, the examples could be misleading if interpreted to be anything more than just a demonstration of a principle. The main emphasis will be to discuss several economic aspects of the design and how the various alternatives interrelate with the cost, and in some cases, how they influence reliability, controllability, or safety. Additional information on pipeline economics can be obtained from References 3 and 55.


### Expected life

The cost of the pipe (materials and installation) is often one of the largest single expenses of a project. Several factors must be considered in arriving at the most economical type of pipe. The initial cost and the life expectancy of a pipeline vary with the type of material and requirements for linings or protective coatings.

**TABLE 2.2    Present-Worth Analysis (cost/ft) Based on 80-yr Life and 10% Simple Interest**

| Pipe | Assumed Useful Life (yr) | Initial Installed Cost/ft ($) | Replacement Cost | | | Present Worth ($) |
|------|------|------|------|------|------|------|
| | | | 20 yr ($) | 40 yr ($) | 60 yr ($) | |
| 1 | 20 | 75 | 11.15 | 1.66 | 0.25 | 88.06 |
| 2 | 40 | 85 | | 1.88 | | 86.88 |
| 3 | 80 | 100 | | | | 100 |

Comparing the cost of several pipes that have different expected lifes must be based on present worth. This requires establishing the expected life of the project, estimating future repair and replacement costs, and adjusting those costs to present day dollar values. The present worth is the sum of the initial cost and the adjusted replacement costs. The pipe with the smallest present worth is presumably the best choice. The uncertainties in this process are related to estimating the useful life of a pipe material accurately, forecasting present and future replacement costs, and projecting inflation and interest rates. It is also necessary to place some value on safety, reliability, and availability.

The data in Table 2.2 demonstrate an 80-year present worth comparison of three pipes with useful lifes of 20, 40, and 80 years, respectively. The present worth of future replacement costs is calculated by

$$\text{Present worth} = \frac{\text{initial cost}}{(1 + i)^n}$$

$n$ = years of useful life

$i$ = net interest rate (interest-inflation)

The comparison, which only considers replacement cost, indicates that pipe 2 is the economical choice based on an net interest rate of 10%.

Other factors which are difficult to include in such an analysis include disruption of service during replacement, disturbing the right-of-way, and repair costs.

### Economic Life

Economic life refers to how long a pipeline will satisfy demands. Such an estimate requires prediction of population growth and industrial expansion. The problem often reduces to a decision of whether to build one large pipeline initially or build a smaller line and add a second later. Reference 3 contains an analysis of pipe diameter, capacity, and cost. The comparison was made by selecting the flow rate so that each pipe had the same head loss (4 ft/1000 ft, with $C_h = 130$). Even though the costs are outdated, the relative cost of the different pipe sizes can be used to demonstrate the procedure. Table 2.3 summarizes the data.

*Example 2.1.* To demonstrate the use of the information in Table 2.3, compare the cost of supplying 30 mgd by one pipe versus using two.

From Table 2.3, a 36-in. pipe can supply 32.6 mgd. The relative cost per foot is $32.6 \cdot 1.27 = \$41.40$. Note that it is necessary to use the capacities listed in Table 2.3, not the actual flow rates.

The 30 mgd can also be supplied by a 30-in. and a 24-in. pipe in parallel. The relative cost per foot would be $20.2 \cdot 1.73 + 11.4 \cdot 2.64 = \$65.04$. The one 36-in. pipe is much cheaper, assuming both pipes are built at the same time.

Now consider that the projected demand is such that the 24-in. pipe can satisfy the demand for 10 years, at which time the 30-in. pipe will be built. Comparing on a present-worth basis using 10% net interest gives a relative cost of $\$43.57$.

The one 36-in. pipe is still slightly more economical, based on the assumptions. The longer the time before the second pipe is needed, the more economical the two pipes will be.

In contemplating such a decision, there are several hydraulic problems that need to be considered. Solving these problems may have an effect on the economics. If the large pipeline is built and present demands are significantly less than design flow, one must be concerned about cavitation,

TABLE 2.3    Variation of Pipe Cost With Diameter[a]

| Pipe diameter (in.) | Relative cCapacity (mgd[b]) | Pipe Unit Cost[c] | Pipe diameter (in.) | Relative Capacity (mgd) | Pipe Unit Cost[c] |
|---|---|---|---|---|---|
| 8 | 0.6 | 18.2 | 30 | 20.2 | 1.73 |
| 12 | 1.8 | 8.09 | 36 | 32.6 | 1.27 |
| 16 | 3.9 | 5.36 | 42 | 48.5 | 1.18 |
| 20 | 7.0 | 3.77 | 48 | 70.0 | 1.00 |
| 24 | 11.4 | 2.64 | | | |

[a]Reference 3. (Courtesy the American Society of Civil Engineers.)
[b]mgd = million gallons per day.
[c]Costs are based on the ratio of the cost/ft of head loss/mgd to the cost of a 48-in. pipe.

sedimentation, air removal, and the possibility of open-channel flow. Cavitation may be important, since at low flow rates there will be little friction loss, and control valves may be required to regulate the flow or pressure. High pressure drops increase the chance of cavitation. Preventing cavitation may significantly increase the cost of control structures.

If the initial system demand results in velocities less than that required to flush air and prevent settling of suspended solids, some provision may be needed for periodically flushing at higher velocities. The velocity must, at least part of the time, be high enough to move the air to air release valves. If not, air pockets can become large enough to reduce the flow or cause transients if they move through the pipe as large air pockets.

At low flow rates, it is also possible to have open-channel flow on steep grades. Such flow will normally be supercritical and terminate with a hydraulic jump. Large quantities of air can be entrained which cause problems by either moving through the pipe or blowing back, causing surges. Also, the higher open channel velocities, the turbulence of the hydraulic jump, and the large amount of air can cause erosion and corrosion of pipes and liners. The possibility of open channel flow can be predicted by calculations and avoided by placing additional valves in the line to maintain positive pressures everywhere in the pipe. This problem is most likely to occur on the downstream side of hills.

For pumped lines, there are additional problems posed by building pipelines for future demands. The variable demand is normally allowed for by installing additional pumps as demand increases. As the flow rate goes up, the pumping head must increase to overcome the additional friction loss. Centrifugal pumps, normally used for such service, produce less flow as the head increases. There is a rather narrow range of head and flow where the pump operates efficiently and free of cavitation or other problems. If the first pump is designed to provide the proper head and flow for initial demands, it will be inadequate for future use when the head increases significantly. This may require replacing the entire unit or at least the impeller and motor.

Another possible solution, if demand is expected to increase rapidly, is to choose the original pump to meet future head requirements and use a discharge control valve, orifice, or other means of maintaining the necessary pressure on the pump until demand increases. The design of the pressure-sustaining device needs to be such that excess cavitation, noise, vibrations, or torque on the valve is avoided. The pressure-reducing equipment and the additional power costs caused by the wasted pumping head must be included in the economic analysis and compared with pump replacement costs.

Table 2.3 demonstrates that the cost per unit volume of transporting a liquid decreases as the pipe size increases. The area of a pipe is proportional to the square of the diameter, so doubling the diameter will increase the capacity by fourfold. In addition, larger pipes have a smaller relative roughness and therefore will have less friction and a higher velocity for a given available head.

The weight per unit length of the pipe, which is directly related to the cost, increases by slightly more than double as the diameter doubles. Installation and transportation costs also increase with size, but the cost per unit volume of water may decrease. The net effect is a decrease in cost per unit volume of liquid with pipe size. This has a major impact on the economic life and the decision of building one large or two smaller pipelines, as discussed in the preceding section.

Once the economic life, and therefore the design flow is established, the most economical pipe diameter can be selected. For a gravity system, the smallest size that will provide the design flow is chosen (with proper provision for aging). Pipes only come in certain sizes. For example, with steel pipe, standard sizes usually come in 2-in. (5.1-cm) increments between 4 and 20 in. diameters. Above 24 in. they usually increment by 6 in. The available pipe sizes vary with type of material. Manufacturers should be consulted for complete information. Because of these large diameter changes, it is seldom possible to choose a single pipe size that just uses up the available head. One alternative is to determine the proper lengths of two pipe diameters that produce the proper head loss. In addition to reducing diameter, it may be possible, or necessary, to change pressure class of pipe. In selecting the pressure class, allowance for future changes in demand and transient protection should be considered.

In long gravity lines with limited head, it may be advisable to compare the cost and practicality of using one or more booster pump stations and reducing the pipe size. This would be mandatory if the available head would not provide a minimum allowable velocity. Pumps are often the solution to future demands since they can be added as needed.

*Pumping Costs.* When pumps are being considered as an alternative, the economic analysis consists of finding the proper combination of pipe diameter and pumps that produces the least annual cost. Example 2.2 demonstrates the various items that must be considered in such analysis.

> *Example 2.2.* A 10,000-ft horizontal pumped pipeline is required to supply 14 cfs. Demands are such that the pumps will operate 5 h/day. Select the most economical pipe diameter based on total annual cost. Assume the following: 1) $C_h = 130$ for all pipes; 2) power costs are $0.07/kW-h and the demand charge is $5/kW/month; 3) 80% efficiency for the pumping plants; 4) the cost of the pumping station is amortized over 20 years at 10%; 5) the cost of the pipeline is amortized over 40 years at 10% net interest.
>
> Values for installed pipe cost (item 1, Table 2.4) were taken from Reference 3 and adjusted for inflation. Head losses are calculated with the Hazen–Williams equation.
>
> $\text{hp} = Q_\gamma H / 550e$, $\text{kW} = 1.341 \text{ hp}$ ($e$ = efficiency of pump and motor)

**TABLE 2.4 Selection of Pipe Diameter and Pump Capacity for Example 2.2**

| | | | | $D$ (in.) | | |
|---|---|---:|---:|---:|---:|
| | | 12 | 16 | 20 | 24 |
| 1. | Total cost of installed pipe ($) | 360,000 | 530,000 | 660,000 | 860,000 |
| 2. | Required pumping head | 776 | 191 | 64.4 | 26.5 |
| 3. | Horsepower (80% efficiency) | 1,540 | 379 | 128 | 52.6 |
| | Required electrical power (kW) | 1,150 | 283 | 95.4 | 39.3 |
| 4. | Annual power cost of pumping ($) | 216,000 | 53,100 | 17,900 | 7,370 |
| 5. | Cost of pumping station ($) | 606,000 | 148,000 | 75,000 | 50,000 |
| 6. | Annual cost of pumping station ($) | 71,200 | 17,400 | 8,800 | 5,860 |
| 7. | Total annual pumping cost (4 + 6) ($) | 286,900 | 70,500 | 26,730 | 13,250 |
| 8. | Annual cost of pipeline amortized over 40 years ($) | 36,800 | 54,200 | 67,500 | 87,920 |
| 9. | Total annual cost (7 + 8) ($) | 323,700 | 124,700 | 94,200 | 101,900 |

Annual power cost = use charge + demand charge =

kW(365 days/yr · 5 h/day)$0.07/kW-h + kW · $5/kW-mo · 12 mo/yr

Annual cost of pumping station or pipeline:

$$= \text{Total cost } \frac{i(1+i)^n}{(1+i)^n - 1}$$

$i$ = net interest, $n$ = years

The total annual cost for the 20-in. pipe is the smallest (row 9 of Table 2.4) so it would be the best choice based on the assumptions. The relationship between pipe cost, pumping cost, and total cost per year is demonstrated graphically in Fig. 2.1.

When filling in the data for Table 2.4, several factors must be considered: 1) Pumping costs should be based on a best guess of actual quantity of water pumped. If the demand increases significantly over the years, a best estimate of the average yearly water should be used rather than the amount based on the design flow. 2) Will the pumps operate at their design point where they are most efficient? To accommodate this, pumped systems should have adequate terminal reservoirs so the pumps only replace used reservoir water and are not required to meet varying system demands. Properly designed, this will allow the pumps to operate near maximum efficiency. If this is not possible (as with many booster pump stations), a lower efficiency must be used to compute power consumption. If both seasonal and daily demands are not handled by storage, pumping costs will increase. 3) Interest rates and power costs are difficult to predict and cloud any conclusion with uncertainty. 4) Electrical power rates normally include a "demand charge" which can have a significant impact on pumping costs. The demand charge is based on the maximum power use during the month, regardless of how long it was used. 5) When considering a smaller-diameter pipe combined with

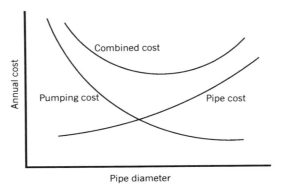

**Fig. 2.1**   Selection of optimum pipe diameter.

higher pumping heads, care must be used not to exceed the pressure rating of the pipe. If this happens, the next higher class of pipe should be used and proper adjustment to the pipe cost should be made.

When the economic life has been reached and demands exceed the capacity of an existing pipeline, the decision must be made to install more storage, booster pumps, or an additional pipeline. If the increased demands are primarily in the peak flow rates, the problem may be solved by installing additional storage. If daily consumption exceeds total supply capability, then the supply must be increased. The economic analysis of such a choice is basically the same as that described in Table 2.4 and Fig. 2.1. The choice depends primarily on the amount of increased capacity required.

If only a modest increase in capacity is required, pumps may be the economic choice. Power consumption increases with approximately the cube of the velocity. This increase in power cost may be excessive, especially if there is a high demand charge. Therefore, large increases of flow by pumping rapidly become uneconomical. The allowable flow increase by pumping is also limited since the pressure class of the pipe cannot be exceeded.

Another variable associated with designing pumping plants is the number of pumps and pumping stations, the technical aspects of which are discussed in Chapter 3. One advantage of multiple pump stations located at different points in the system is that lower pressure class pipe can be used, which may result in significant savings. Another is that lower head pumps are less subject to wear, especially if there is sediment in the water. A third is that all pump stations may not be required initially, reducing the initial investment.

On the negative side, one must consider the overall cost of multiple pump stations, including the cost of getting power to the various sites, access roads, and the increased complexity and cost of equipment and controls for operating the pumps and for transient protection.

A decision must also be made about the number of pumps installed at each site. For flexibility and reliability, multiple units usually are installed. If there is a fairly constant demand, a typical installation would consist of three identical units, any two of which can supply the design flow. If there is a large variation in flow that is not satisfied by storage tanks or reservoirs, several pumps of different capacity may be installed. In doing this it is necessary to match properly the head capability of any pumps that will operate simultaneously.

## Storage

The purposes of storage tanks or reservoirs are to: 1) supply water when there is a temporary interruption of flow from the transmission main, 2) provide supplemental water during peak periods, 3) sectionalize the pipe to reduce mean pressures and transient pressures, 4) maintain pressure (elevated storage), and 5) simplify control.

The reliability of a system, although tempered by economics, is more a social or political decision. If an uninterrupted supply is desired, storage facilities must be included, regardless of cost. On the other hand, if the purpose of considering storage is to reduce the capacity of the transmission main so it does not have to supply the peak demand, then an economic evaluation is necessary. This is similar to the pumping cost versus pipe diameter study. First, calculate the cost of a pipeline that would supply the peak demand and require no storage. Next, calculate the cost of one or two smaller pipes or combinations of pipes and determine the amount of storage required to supply supplemental water during high flows. This produces a pipe cost versus diameter curve similar to that shown in Fig. 2.1. Next, calculate the cost of the storage facilities. This will produce a line similar to the pumping cost line in Fig. 2.1. Adding the two gives the combined cost and will identify the least cost solution.

Storage also has a significant impact on the control structures, pumping plants, and general operation of the pipeline. If there is adequate storage, large fluctuations in demand can be tolerated. Any mismatch in supply and demand is made up by increase or decrease in storage. Valves in the transmission main require only infrequent adjustments to maintain storage. Pumps can operate for long periods at their design point. They are normally activated by level controls at the storage tank and not by fluctuations in demand.

If there is no storage, the system must be able to provide continuous fine adjustment of the flow within safe pressure limits. For gravity systems, this requires pressure-regulating valves at selected points that react automatically. For pumped systems, the variations in flow will cause the pumps to operate both below and above their design point where power consumption is high, efficiency is low, or where there is more chance of operational problems.

For systems where there are large elevation changes, intermediate reservoirs can be used to limit the pressure, make controls simpler, and reduce transients.

Consider Fig. 2.2, which is a profile of a major pipeline supplying southern California (Reference 84). Water is conveyed from the supply reservoir at 1500 ft (457 m) elevation to a lower reservoir, elevation 345 ft (105 m). If a single control structure were installed at the lower reservoir, much of the pipeline would be subject to shutoff pressures near 650 psi (4.48 MPa). This would require a valve structure with a pressure drop of 500 psi (3.45 MPa) at shutoff. Both of these produce undesirable situations.

One solution is to use intermediate control structures, preferably with small storage tanks. Without open water surfaces, it would be necessary to depend on the control valves and pressure relief valves to prevent the entire

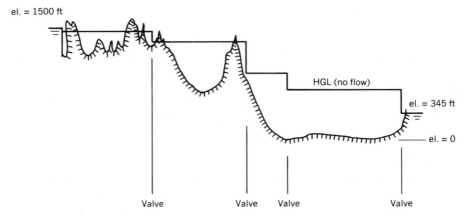

**Fig. 2.2**   Profile for Foothill feeder pipe.

line from reaching maximum shutoff pressure. The control structures should be designed so that the required pressure drop does not create excessive cavitation problems and so that lower pressure pipe can be used. The number and cost of such installations could be offset by the savings in pipe cost and improved operational safety.

### Control Structures and Equipment

During the feasibility study, only a general design has been completed, so a detailed analysis of all hydraulic problems and their solution is not available. Even so, it is necessary to anticipate the need for special facilities or equipment and make some provision in the economic analysis. Such predictions will be refined as the design continues. Considerations would include the need for water treatment, eliminating excessive cavitation, and controlling transient problems. If water treatment is required, it is often a major cost factor and may be handled as a totally separate design project. Water treatment is not included in this text.

The control or prevention of hydraulic transients can be a significant cost factor for some pipelines. An experienced designer can usually identify situations that will require special attention. Provision should be made for the cost of the detailed analysis, design, and construction required to control transients.

Some attention should be given to costs associated with winterizing, highway crossings, special geologic or topographic problems, possible testing or model studies, and any other items that would have a significant influence on the cost, reliability, or safety of the project.

## 2.3  PIPE MATERIALS

The materials used for pressure pipe include cast iron, ductile iron, steel, reinforced concrete, plastic, and fiber glass. For low-pressure applications, one can also select unreinforced concrete, corrugated steel, or a new product called spiral rib (65). The choice of a material for a given application depends on hydraulic roughness, pipe size and pressure requirements, resistance to internal and external corrosion, ease of handling and installing, the environment where the pipe will be installed, useful life, and economics.

The most basic requirement for a pipe material is that the pipe must be able to withstand the maximum internal pressure. In addition to considering the positive pressure rating of a pipe, it is necessary to determine if the pipe will be subjected to any negative pressures. Negative pressures can collapse a pipe unless the condition exists only momentarily, such as during a transient. If a pipe is buried, the soil and groundwater pressure plus live loads may be sufficient to cause collapse even if the pressure inside the pipe is positive. For details, consult Reference 79.

Selection of wall thickness for larger pipes often depends more on collapse pressure and handling loads than it does on burst pressure. A thin-walled, large-diameter pipe may be adequate for resisting relatively high internal pressures, but may collapse under negative internal pressure or while being handled. This problem increases with pipe diameter and especially affects plastic and thin-walled steel pipe. Some pipes are made to resist full internal vacuum. For example, concrete generally has thick enough walls that it is very resistant to collapse.

The resistance of the different materials to impact loads and forces caused by poor bedding conditions is important. Steel, ductile iron, and cast iron are resistant to impact load caused by external forces or transients. Less care is needed in handling or burying these pipes. Concrete, plastic, corrugated metal, and spiral rib are more susceptible to damage during handling, shipping, and installation. For these materials, it is necessary to provide proper backfill and bedding conditions to avoid damage due to external loads. The available specifications on the different pipe materials elaborate on this and should be carefully studied.

One of the real advantages of concrete and various plastic materials is their corrosion resistance. The metallic pipes usually need interior linings and exterior coverings. Cement mortar, epoxy, and coal tar are commonly used for such purposes. Some linings decrease the internal diameter or require a thicker wall to prevent deflections that could cause damage to the linings. For some linings, field application at connections is necessary, which increases the cost.

Ease of transporting, installing, and method of connecting the pipe sections have a significant effect on total cost. Plastic, corrugated metal, and spiral rib have an advantage since they are lightweight.

Methods used to couple the pipe sections depend on the pipe material. A variety of connectors are available for metallic pipes, including welding, bell and spigot, flange, mechanical fasteners, and ball and socket for large deflections. Concrete pipe is usually joined with confined O-ring gaskets and steel or concrete retainers. Plastic pipe comes with bell and spigot joints, glue joints, flanges, and mechanical fasteners, and in small sizes, may be threaded. Corrugated metal and spiral rib pipe use various types of band couplers.

## 2.4  PRESSURE CLASS OF PIPE

A fundamental step in proper pipe design is selection of the pressure class of the pipe. A thorough discussion of specific procedures for each of the common pipe materials is beyond the scope of this text. Guidelines for each type of material are available from the American Water Works Association (AWWA), the American Society for Testing and Materials (ASTM), American National Standards Institute (ANSI), Plastic Pipe Institute (PPI), Federal Specifications (FED), and specifications from the pipe manufacturers. Such specifications are available for all of the common types of pipe materials. Individuals involved in designing pipelines should obtain and study specifications for the pipe materials they are considering. The principles involved in selecting a pressure class for the various pipe materials are complex, so only general guidelines will be given. Those interested in details should consult design handbooks for each pipe material.

The primary considerations in selecting a pipe pressure class are: 1) maximum internal operating pressure; 2) surge or transient pressures; 3) variation of pipe properties with temperature or long-term loading effects (especially for plastic pipe; 4) damage that could result from handling, shipping, installing, or reduction in strength due to chemical attack or other aging factors; and 5) external loads, including both earth loads and live loads. Each of these will be briefly discussed.

Most pipe is hydrostatically pressure tested for quality control purposes. The test varies with the type of material. These pressure tests are conducted at the manufacturing plant on pipe that is new and in ideal condition with no external load applied. The pressure testing is normally carried out in a short time with the pressure slowly and evenly increased to the set limit or until it bursts. For the nondestructive tests, the pressure is held constant for a short time and then relieved. For example, PVC pipe for municipal water mains manufactured in accordance with AWWA Standard C900 is hydrostatically proof-tested at four times its rated pressure class.

There can be misunderstandings regarding these hydrostatic test pressures. The inexperienced engineer may be tempted to conclude that as long as the maximum operating pressure in the system never exceeds the hydrostatic test pressure, the pipe will not rupture. This is wrong, since the five factors listed above reduce the safe operating pressure of the pipe and must be considered. Another design procedure that is sometimes used is to select a pipe pressure class so that the maximum system pressure is less than the pressure rating of the pipe. However, this procedure may not always be adequate if there are high transient pressures or large external loads. Each type of pipe is designed differently, so the following guidelines given are general guidelines only.

The maximum operating pressure in a gravity flow system is simply the reservoir shutoff head. For a pumped system, it can be the pump shutoff head. One may use these values or elect to be less conservative and use the actual pressures for normal flow.

Surge or transient pressures are difficult to predict. They depend on the specific design and operation procedures of each pipe system. Some standards give general guidelines that can be followed if a transient analysis is not made. To demonstrate the potential problem, consider the following example. Engineering calculations of anticipated waterhammer pressures in a pipe distribution system for a particular installation predicted that the maximum pressure, including waterhammer, would be under 125 psi. After several ruptures of the pipe occurred, field tests were conducted. The tests showed that pressures as high as 540 psi were generated by rapid valve closure. For this case, use of general guidelines was inadequate. The later chapters of this book discuss how to make a transient analysis of piping systems. It is not unusual to experience transient pressures that exceed the general guidelines.

For plastic pipe, temperature has a significant effect on its strength. Raising the temperature from 70 to 110°F reduces the pressure capacity of PVC pipe to about half of its original strength. There is also a long-term time effect that reduces the burst pressure of plastic pipe. This is accounted for with a safety factor.

Other factors that can reduce the safe operating pressure of a pipe include handling, shipping, installing, chemical attack, and aging. These are often impossible to predict and are allowed for with a safety factor.

The preceding design factors can be combined into a term called the "internal hydrostatic design pressure." This is equal to

$$\text{Internal design pressure} = (P_o + P_s)SF \qquad (2.1)$$

where $P_o$ is the maximum steady-state operating pressure, $P_s$ is the surge or waterhammer pressure, and $SF$ the safety factor applied to take care of the unknowns just enumerated. The safety factor is usually recommended to be at least two or three. There is no established guideline in the specifications about whether to include unusual transients in the $P_s$ or provide for them in the safety factor. Transient analysis is a new field and those writing the specifications are not schooled in recent techniques. They generally just recommend that experts in the field be consulted to perform the analysis to provide $P_s$.

The criteria discussed thus far have ignored the influence of externally applied loads. In addition to pressure tests on pipe, external loads are applied to test the pipes resistant to external forces. For plastic pipe, the load is applied until it deforms a predetermined amount. If it does not split, crack, or break, it passes the test. These tests are conducted with no internal pressure. It is therefore clear that if the external load in the pipe is high enough, a pipe can break with little or no internally applied pressure. This makes it necessary to consider the influence of external loading during the selection of the pipe pressure class.

The magnitude of the external load depends on the diameter of the pipe, pipe material, the trench width, the depth of cover, the specific weight of the soil, the degree of soil saturation, the type of backfill material, the method used to backfill, the degree of compaction, and live loads. The earth load increases with width and depth of the trench. The live load reduces with depth of cover. The cumulative effect of all these sources of external loading requires considerable study and analysis. Those interested in the details should consult Reference 79. Because of the complexity of this analysis, it is usual to assume that the safety factor is adequate to account for external loads as well as the other factors already mentioned. This cannot guarantee trouble-free operation. Problems may occur in a system with the right combination of factors. For example, if there was some pipe damage due to shipping or installing, plus some unusual transients, if the live or dead loads are unusually high, etc., the safety factor may not be adequate to protect the pipe. In other words, the safety factor cannot replace good engineering judgment and calculations. In many applications, it is adequate to select the pipe pressure class so that the maximum system pressure (the normal system pressure plus normal transient pressures) is less than the pipe pressure rating of the pipe. This leaves the entire safety factor for unusual transients, live and dead loads, and uncontrollable factors such as damage due to transportation, installation, and aging.

## 2.5  HYDRAULIC DESIGN

The emphasis of this text is on the design, installation, and operation of water mains, which are the major arteries of pipe systems. They can be subdivided into transmission mains and distribution mains and may be the most expensive and most important part of a system. Reliability is of major concern. Interruption of service affects large areas and is undesirable. Because these mains are usually long, the chance of damaging transients is a problem. Control valves are often required to regulate a large flow range. At low flow rates, large pressure drops are often necessary and cavitation must be considered.

Although these same problems exist in pipe distribution networks, they are generally less severe. For example, the short interconnected pipes of a network effectively dissipate transients and reduce the need for detailed analysis and transient control equipment. Cavitation is usually less of a problem, since there is usually fairly high line pressure and only modest pressure drops required at valves. There are, of course, situations in which transients and cavitation are important. The designer should be able to identify these situations for such systems and use the same principles developed for the main lines to solve the problems.

In this section, examples of designing simple gravity and pumped pipelines are discussed. The detailed procedures for analyzing branching pipes and complex networks are not presented. The reader is referred to other sources for details.

Before proceeding with examples, a few comments will be made about units. There are several groups involved in designing pipelines: mechanical engineers, civil engineers, waterworks engineers, and suppliers of equipment and pipe. It is somewhat unfortunate that different systems of units are used by these groups. Another degree of complexity is added by the transition to the metric or international system (SI) of units.  In the waterworks field, pipeline capacity is usually expressed in million gallons per day (mgd), pump capacity in gallons per minute (gpm), and storage volume in millions of gallons. Civil engineers typically use cubic feet per second (cfs) for flow rate and pump capacity and acre feet for storage volume. Conversion factors for the various systems of units are given in Table A.1. In the text, all three systems are used.

### Gravity system

The principles involved in selecting the correct pipe diameters were discussed in Section 1.4. Application of those principles to a gravity system is demonstrated by the following examples.

*Example 2.3.* A 10,000-ft pipeline is required to transport 15 mgd from a supply reservoir to a storage reservoir. The supply reservoir water level varies seasonally from elevation 2760 to 2790 ft. The storage reservoir is at elevation 2600 ft. Find the most economical of pipe size.

The design head must be the minimum available to ensure that the required discharge can always be supplied, so:

$$H_f = 2760 - 2600 = 160 \text{ ft}$$

Since the pipe is long and no flow regulation is required, minor losses can be ignored. Use the Hazen–Williams equation (Eq. 1.28) to calculate friction losses with $C_h = 140$ (obtain from pipe manufacturer).

$$V = 1.318 C_h R^{0.63} S^{0.54}$$

Assume that a 150-psi pressure class of pipe is selected. Inside pipe dimensions are obtained from manufacturer's published data.

$$Q = 15 \text{ mgd} = 15 \cdot 1.547 = 23.2 \text{ cfs}$$

The problem will be solved by selecting several different pipe sizes, and calculating the friction slope $S$ from Eq. 1.28a and the corresponding $h_f = SL$.

| $d$ (Inside Dia.) (in.) | $V$ (fps) | $R = d/4$ (ft) | $S$ | $h_f$ (ft) |
|---|---|---|---|---|
| 18 | 13.1 | 0.375 | 0.0234 | 234 |
| 20 | 10.6 | 0.417 | 0.0140 | 140 |
| 24 | 7.38 | 0.500 | 0.00578 | 57.8 |

The 20-in. pipe will work, since its head loss is only 140 ft compared with the minimum available of 160 ft. This allows 20 ft for aging and minor losses.

To optimize the design, one could use some 20-in. and some 18-in. pipe so that the total loss just equals 160 ft. Let $x$ be the unknown length of 20-in. pipe:

$$0.0140x + 0.0234(10{,}000 - x) = 160, \qquad x = 7{,}872 \text{ ft of 20-in. pipe}$$

To demonstrate the influence of pipeline profile on the design, consider the problem shown in Fig. 2.3. The problem will be solved this time with the Darcy equation. The steps in the solution were outlined in Section 1.4.

*Example 2.4.* It is desired to use steel pipe with a 3/16-in. coal-tar enamel lining for the system shown in Fig. 2.3. The roughness recommended by the pipe manufacturer is assumed to be 0.0002 ft. For best accuracy, use actual inside pipe diameters (ID). For steel pipe above

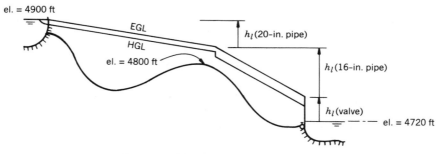

**Fig. 2.3**  Pipeline for Example 2.5.

12 in, the nominal dimensions refer to outside diameter (OD). Use a wall thickness of 3/8 in. so ID = OD − 2·3/8 − 2·3/16 = OD − 1.125 in. The design data are: $H_l = 180$ ft, $Q = 15$ cfs, $L = 8000$ ft total (5000 ft to top of hill), $\nu = 1.66 \times 10^{-5}$ ft²/s (Table 1.1 at $T = 40°F$)

For an assumed $f = 0.015$, Eq. 1.25 gives $d = 1.31$ ft or 15.66 in. (ID). The next larger available pipe size would be 18 in. so $V = 15/1.55 = 9.66$ fps, Re $= 8.2 \times 10^5$, $k = 0.0002/1.41 = 0.00142$, and from Eq. 1.26, $a = 0.0129$, $b = 1.78$, $c = 0.494$, so $f = 0.0150$, and from Eq. 1.25 $h_f = 122$ ft. Since this is less than the available head of 180 ft, one could either use all 18-in. pipe or part 18- and part 16-in. pipe.

In making the pipe diameter selection, consider the pressure at the top of the hill (el. = 4800 ft, $L = 5000$ ft). It is necessary to maintain positive pressure at all locations to prevent air and contaminated groundwater from entering the pipe, to allow air-release valves to function properly, and to prevent pipe collapse.

Calculate the head at the hill with 18-in. pipe installed upstream. Applying the energy equation from the reservoir to the top of the hill (Eq. 1.24),

$$4900 = 4800 + \left(1 + \frac{fL}{d}\right)\frac{V_2}{2g} + P/\gamma$$

$$P = 62.4\left[\frac{100 - 12.42(1 + 85.1)}{2g}\right]\bigg/144 = -45.7 \text{ psi}$$

Since this is below vapor pressure, the pressure at the hill will be at vapor pressure. To prevent negative pressures, use a larger pipe. If a 20-in. pipe is used upstream from the hill, the pressure at the top of the hill will be 23.7 psi.

Downstream from the hill, use 16-in. pipe to increase friction losses. With 3000 ft of 16-in. pipe the head loss is 84.8 ft. The 5000 ft of 20-in. pipe will have a head loss of 44.0 ft.

The head drop required by a control valve to limit the flow to the required 15 cfs would be: $\Delta H = (4900 - 4720) - (84.8 + 44.0) = 51.2$ ft.

The design of control structures to produce head loss without cavitation or transients is discussed in later chapters.

## Pumped System

In Example 2.3, the only part that economics played was in realizing that smaller pipe costs less. When designing pumped lines it is necessary to balance pump installation and operating costs against pipe costs. This is demonstrated in the next example.

*Example 2.5.* Use the same input data as given in example 2.3, but assume it is necessary to pump from the lower to the upper reservoir. Determine the economical pipe diameter. The most demanding case is pumping against maximum water level in the upper reservoir. The design flow and head would be based on the average or most frequently expected elevation difference. Assume a design flow and head of 15 mgd at 175 ft (2775 − 2600) of head. One must also know the total quantity of water pumped annually and accurate power cost data. Assume the total annual pumping requirement is 5000 AF (acre feet), or 1630 million gallons. The calculations are summarized in the table.

Choose a pipe diameter and find the friction head loss (col. 2) for each pipe at design flow (get from Example 2.3).The average pump head = $175 + h_f$.

Find the water horsepower (hp = $Q\gamma H/550$ (col. 4) required to pump 23.2 cfs (15 mgd) at the average pumping head (col. 3). (Average elevation head plus friction loss.)

The average electrical power consumed (col. 5) is calculated by

kW = 0.746 · hp/0.85 (85% assumed efficiency of pump and motor)

| $d$ (in.) | $h_f$ (ft) | Av. Pump Head (ft) | Water (hp) | kW | Yearly Power (kW-h) | Yearly Power Cost ($) |
|---|---|---|---|---|---|---|
| 1 | 2 | 3 | 4 | 5 | 6 | 7 |
| 18 | 234 | 409 | 1,076 | 944 | $2.47 \times 10^6$ | 311,000 |
| 20 | 140 | 315 | 829 | 727 | $1.90 \times 10^6$ | 239,000 |
| 24 | 57.8 | 233 | 612 | 537 | $1.40 \times 10^6$ | 176,000 |
| 30 | 19.5 | 194 | 511 | 448 | $1.17 \times 10^6$ | 147,000 |

Calculate the total pumping time and the annual kilowatt-hours (kW-h) of electricity used (col. 6). Pumping time = 1,630 million gallons/15 mgd = 109 days; kW-h = col. 5 · 109 days · 24 h/day.

Knowing the cost of power per kW-h (assumed $0.08/kW-h) and the demand charge (assume $10.00 per kW per month), the yearly pumping cost (col. 7) is calculated by

Yearly cost = kW-h · 0.08 + 10 · kW · 12.

To complete the example, one would have to estimate the cost of each of four pump installations and the pumps' annual cost based on present worth to arrive at the total annual pumping cost, as was done in Example 2.2.

**Minor losses**

For short systems containing many pipe fittings and valves, minor losses become very important. This is demonstrated in the next example.

*Example 2.6.* The problem consists of selecting a pump that will supply $0.8 \text{ m}^3/\text{s}$ from a lower reservoir (el. 1100 m) to an upper reservoir (el. 1175 m.). Assume that the piping configuration has been set by constraints at the site that require a complex piping configuration. Loss coefficients will be taken from Table 1.4. For more complete values see Reference 37.

| Loss Coefficients in Suction Piping (610-mm dia.) | $K_l$ |
|---|---|
| Flush pipe inlet | 0.50 |
| 4–90 short radius elbows | $4 \cdot 0.24$ |
| 3–30 mitered elbows | $3 \cdot 0.15$ |
| 15 m of 610-mm pipe ($f = 0.014$) | $fL/d = 0.34$ |

| Loss Coefficients in Discharge Piping (408-mm dia.) | $K_l$ |
|---|---|
| Check valve | 0.8 |
| 2 butterfly valves (full open) | $2 \cdot 0.4$ |
| 3 tees | $3 \cdot 0.6$ |
| 6–90 long radius elbows | $6 \cdot 0.19$ |
| Exit | 1.0 |
| 6.1 m of 408 mm pipe ($f = 0.016$) | $fL/d = 0.24$ |

Applying the energy equation (Eq. 1.24) between the two reservoirs,

$$1000 + H_p = 1175 + \left(\frac{fL}{d} + \Sigma K_l\right)_s \left(\frac{V_s^2}{2g}\right) + \left(\frac{fL}{d} + \Sigma K_l\right)_d \left(\frac{V_d^2}{2g}\right)$$

For the suction side: $V_s = 2.74 \text{ m/s}$, $g = 9.81 \text{ m/s}^2$, $\Sigma K_l = 1.91$, $fL/d = 0.34$, $V_s^2/2g = 0.382$.

For the discharge side: $V_d = 6.11 \text{ m/s}$, $\Sigma K_l = 5.54$, $fL/d = 0.24$, $V_d^2/2g = 1.90$.

Solving the equation in step 3 gives $H_p$ (pump head) $= 86.9$ m.

If minor losses were ignored the pumping head would be $H_p = 75 + 0.382(0.34) + 1.90(0.24) = 75.6$ m. This would result in a significant error in selecting the pump.

There are situations in which minor losses become very important because they affect the power generated by a turbine or consumed by a pump. Consider the preceding example with the short discharge line. The minor losses increased the pumping head by 11.3 m, or 15%. The increase in power cost for the life of the project would be significant.

**Branching Systems**

When the flow divides into two or more branches, the hydraulic analysis becomes slightly more complex. A-trial-and-error solution becomes necessary. Consider the example shown in Fig. 2.4. With three interconnected reservoirs at different elevations, it is not apparent whether flow will be into or out of reservoir 2. Assume that the pipe diameters, lengths, and roughness are all specified. The principles involved in the solution are: 1) conservation of mass (continuity equation—Eq. 1.15): $Q1 + Q2 + Q3 = 0$; and 2) the energy equation which specifies that the head in all three pipes at the junction is the same.

Solving for the three discharges and the head at the junction requires an assumed value of the pressure at the junction and four equations. These are the continuity equation and the energy equation applied to each of the three pipes. The accuracy of the assumed head is checked by substituting the three flow rates determined from the three energy equations. into the continuity equation. If the flow into the junction is greater than that leaving, the assumed head at the junction must be increased. This process is continued until the continuity equation is satisfied. It can also be solved by assuming the discharges and checking the head loss in each pipe. The discharges are adjusted until the head losses provide the same head at the junction.

For the simple system shown in Fig. 2.4, solution by a hand calculator is easy. However, as additional pipes are added or if closed loops are created, the system rapidly becomes too time-consuming to solve by this way.

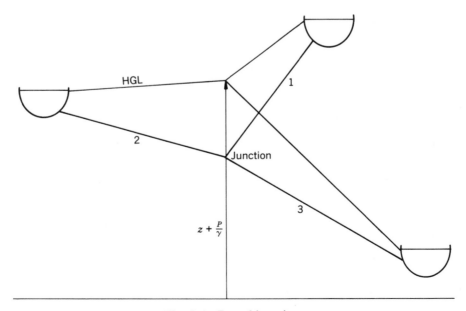

**Fig. 2.4**  Branching pipes.

**Network Analysis**

The principles involved in solving complex pipe systems are the same as those used for the branching system. The solution becomes more time-consuming with increased complexity of the system. All such solutions are now done with computers. There are two basic principles involved in analyzing networks: 1) the algebraic sum of the head loss of all pipes around any closed loop must equal zero. and 2) the algebraic sum of flow at any junction is zero. Implied in item one is the fact that the pressure in all pipes at a junction is the same. This assumes that minor losses at the pipe junction are negligible.

If network analysis is used for design, additional constraints must be placed, such as minimum pressure at junctions or minimum flow in pipes. This adds another dimension to the complexity of the analysis because pipe diameters and the number of cross connections can vary.

There are three general approaches to solving networks: 1) where the flows are the unknowns, 2) where the head at each junction are the unknowns, and 3) where the flow correction is the unknown. The Hardy Cross method (14) was the first systematic method used. Its main limitation is that it solves the multiple equations one at a time rather than simultaneously. Two more advanced methods which use simultaneous solutions are the linear method and Newton–Raphson method (28, 86).

**PROBLEMS**

**2.1.** Find the most economical size of pipe in Table 2.2 based on 15% interest rate and a 10% inflation.

**2.2.** For Example 2.1, assume the 30-in. pipe is built immediately. Find the number of years delay required before building the 24-in. pipe so that the present worth of the two pipes is the same as one 36-in.

*Answer:*   16 yr

**2.3.** In Example 2.2, which pipe would be the most economical if the power cost was \$0.10/kW/h and the demand charge \$10/kW/mo?

**2.4.** When a pumped pipeline is operated at $Q < Q_{des}$, what problems do you encounter with pumping? How can they be solved?

**2.5.** Discuss the hydraulic and economic advantages and disadvantages of using intermediate reservoirs or pressure reducing valves in long pipelines with high gravity heads.

**2.6.** Discuss the advantages of one large pipeline versus two smaller parallel lines.

**2.7.** When water demand exceeds capacity of the pipeline, what are the alternatives?

**2.8.** In a long pumped line, what are the advantages and disadvantages of one booster pump versus several spaced along the line?

**2.9.** At a pump installation, what are the advantages of multiple pumps in parallel versus one pump?

**2.10.** Discuss how storage influences safety, general hydraulics, pump operation, economics, and reliability.

**2.11.** For Example 2.3, find the minimum pipe size if $C_h = 120$. How much head loss would be required by a valve to get the required flow if the reservoir is at low water level?

*Answer*:   83 ft

**2.12.** Calculate the yearly power cost for Example 2.5 assuming the pipe is half 20 in. and half 24 in.

*Answer*:   $208,000

**2.13.** For the branching pipe system shown in Fig. 2.4, calculate the flow in each pipe using the following data:

| Pipe No. | Reservoir Elevation | Pipe Diameter | Length | $C_h$ |
|----------|---------------------|---------------|--------|-------|
| 1 | 1500 | 16 | 2000 | 130 |
| 2 | 1450 | 10 | 600 | 120 |
| 3 | 1400 | 8 | 1000 | 120 |

*Answer*:   $Q = 10.6, 6.08, 4.48$ cfs

# 3

# PUMPS

This chapter discusses the use of centrifugal pumps and their relationship to the hydraulic characteristics of the piping system. Pump characteristics covered include total dynamic head, efficiency, horsepower, net positive suction head, specific speed, suction specific speed, and similarity laws. Examples are given for selecting single pumps, parallel pumps, and series pumps. The hydraulics of pumping pits, suction piping, cavitation, transients, and problems related to starting and stopping pumps are discussed. There is no discussion of theory of pump design or maintenance.

Pumps can be classified into two groups which generally describe how energy is transmitted to the fluid: dynamic and displacement. The two subclassifications of displacement pumps are reciprocating and rotary. Since these are seldom used in water conveyance systems as the principal pumps, they will not be discussed. Those interested in such pumps should consult Reference 30.

The most common type of dynamic pump, and the type discussed in this chapter, is the *centrifugal* pump. This classification is subdivided according to head and discharge characteristics into: *axial*-flow (low head and high discharge), *mixed-flow* (moderate head and moderate discharge), and *radial-flow* (high head and low discharge). These classifications are more accurately defined by the specific speed of the machine (to be defined later), which is a function of head, discharge, and rotational speed. The axial, mixed, and radial classifications can be further subdivided into single- or multiple-stage; single- or double-suction (mixed flow only); open or closed-impeller; self-priming or nonpriming; fixed- or variable-pitch; and fixed- or variable-speed. Pumps can also be classified according to their installation or physical orientation: wet-pit, dry-pit, and horizontal or vertical.

## 3.1  PUMP HYDRAULICS

Selecting a pump for a particular service requires matching the system requirements and pump capabilities. Chapters 1 and 2 provided the tools for analyzing the head–discharge requirements of a system. The process consists of applying the energy equation and evaluating the pumping head required to overcome the elevation difference (static lift) and the friction plus minor losses. For a pump supplying water between two reservoirs, the pump head required to produce a given discharge can be expressed as

$$H_p = \Delta z + H_f \tag{3.1a}$$

or

$$H_p = \Delta z + CQ^2 \tag{3.1b}$$

in which the constant $C = \Sigma(fL/2dgA^2 + K_l/2gA^2)$

Figure 3.1 is a graphical representation of Eq. 3.1 showing the general shape of a system head curve. The curve shown is for a system having a relatively large elevation change and significant friction losses. The shape of the system head curve depends on the relative magnitudes of the elevation change versus friction losses.

Before a pump can be selected for a particular application, information must be provided from which the system head curve can be generated. If the elevation of either reservoir is a variable, then there is not a single curve but a family of curves corresponding to the various reservoir elevations. In addition to supplying the system head curve information, it is necessary that the desired operating range be identified before any pump or combination of pumps can be selected.

With the system head and discharge characteristics and the approximate operating point determined, selection of a pump is possible. Proper selection of a pump requires that it not only provide the required head and discharge, but that it operate near its rated conditions, which is its best efficiency point (bep), and function free of cavitation, vibrations, and any

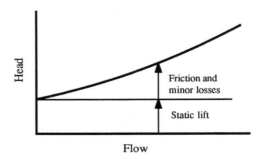

**Fig. 3.1**  Typical system head curve.

other undesirable characteristics. The entire piping system should also be analyzed from the standpoint of any hydraulic transients which may be generated by start-up, shutdown, and any other normal or abnormal changes in the flow.

**Total Dynamic Head**

Before discussing pump characteristics, several parameters used to describe pump performance should be defined. The pump head $H_p$ discussed in connection with Eq. 1.24 is usually referred to as the total dynamic head of the pump. It is the change in the energy grade line at the pump. An explanation of what it represents is obtained by considering how one would experimentally measure the total dynamic head for a pump installed in a pipeline. Assume that the pump is installed with a straight section of suction pipe and a straight section of discharge pipe both of sufficient length to develop uniform flow. Assume further that piezometer taps are installed in the suction pipe several diameters upstream from the pump inlet and in the discharge pipe at a sufficient distance that uniform flow exists at the piezometers. Rewriting Eq. 1.24 (neglecting $H_t$, the turbine head) and solving for total dynamic head $H_p$ produces:

$$H_p = \frac{V_2^2}{2g} - \frac{V_1^2}{2g} + P_2/\gamma - \frac{P_1}{\gamma} + z_2 - z_1 + H_f \qquad (3.2)$$

This equation represents the total increase in energy created by the pump, expressed in feet or meters of fluid between section 1 and 2. This total dynamic head includes 1) any increase in dynamic head ($V_2^2/2g - V_1^2/2g$) created by having the discharge pipe smaller than the suction pipe, 2) increase in the pressure head ($P_2/\gamma - P_1/\gamma$), 3) any elevation change between the suction and discharge piping ($z_2 - z_1$), and 4) the pipe friction losses which occur between the piezometers in the suction and discharge pipes ($H_f$). If $H_p$ is calculated using $P_1$ and $P_2$ from the hydraulic grade line projected back to the suction and discharge sides of the pump, then $H_f$ in Eq. 3.2 would be zero. If the pump is supplying water between two reservoirs and points 1 and 2 are selected at the surface of the reservoirs, the equation reduces to $H_p = z_2 - z_1 + H_f$. In this case, $H_f$ includes all friction and minor losses for the entire system.

Equation 3.1 shows that the pump head is related to the square of the velocity or discharge. If there is a valve in the discharge piping, it is possible to vary the flow through the pump, which results in a corresponding change of pump head. By measuring the total dynamic head at different discharges, one generates what is referred to as a pump rating curve. Typical rating curves (or characteristic curves) for constant speed centrifugal pumps are shown in Fig. 3.2. Different characteristic curves can be generated by

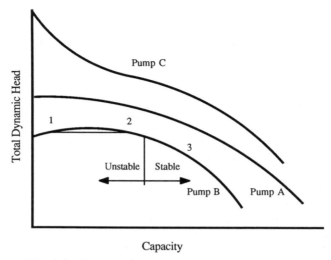

**Fig. 3.2** Pump rating curves for centrifugal pumps.

changing the speed of the pump or impeller diameter. At zero flow, the total dynamic head is referred to as the shutoff head.

Curve A in Fig. 3.2 is called a stable or normal rising pump characteristic. As the flow is reduced, the head continually increases. Curve B is an example of an unstable or drooping characteristic because below some flow the head reduces as the flow decreases. Such a pump is unstable because at low discharges the flow can oscillate between two values. Points 1 and 2 on curve B of Fig. 3.2 represent two such flow rates. When the pump tries to operate at a flow below that corresponding to point 3 on curve B, the flow can be unstable. This results in fluctuations in the electrical load and creates pressure surges in the pipeline. Such situations should be avoided either by selecting pumps that have stable characteristics or by making provisions that prevent an unstable pump from operating near its unstable zone. The third type of characteristic shown in Fig. 3.2 is called a steeply rising characteristic (pump C). This type of characteristic is useful when the system pressure varies significantly.

A sample set of pump characteristic curves for a centrifugal pump is shown in Fig. 3.3. Data are shown for three impeller diameters, labeled as curves A, B, and C. The figure includes information on head, flow, efficiency, net positive suction head, and brake horsepower. Each of these terms will be discussed in the next section. The best efficiency point (bep) or normal operating point would be near the middle of the area of 85% efficiency.

For computational purposes, especially for computer applications, it is convenient to express the head-flow part of the pump rating curve as an

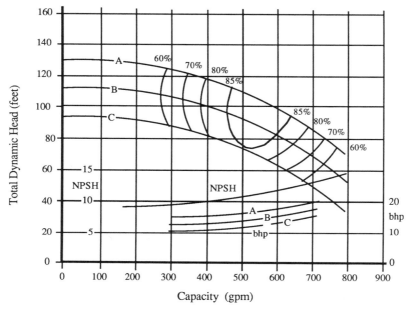

**Fig. 3.3**  Pump rating curve for low specific speed pump.

equation. For a centrifugal pump with a normal characteristic curve, operating near the design point, the head can be related to the discharge by

$$H_p = H_0 - C1Q - C2Q^2 \qquad (3.3)$$

in which $H_0$ is the shutoff head (head at $Q = 0$), and $C1$ and $C2$ are constants evaluated for each pump curve. The constants are evaluated by substituting $H_0$ and two sets of $Q$ and $H$ values scaled from the curve near the design point into Eq. 3.3 and solving simultaneously for $C1$ and $C2$. This form of the equation is used in Chapter 9 for transient analysis of pumps. The equation can also be solved simultaneously with the system equation to evaluate the pumps flow rate.

### Mechanical and Electrical Power

The horsepower delivered by a pump to the fluid, referred to as water horsepower, is calculated by:

$$\text{whp} = Q\gamma \, \frac{H_p}{550} \qquad (3.4)$$

in which $Q$ is the flow rate in cubic feet per second, $\gamma$ is the specific weight of water (Table 1.1), and $H_p$ is the total dynamic head in feet, evaluated from Eq. 3.2.

When using $Q$ in gpm and specific gravity (sg), water horsepower is calculated by

$$\text{whp} = \text{sg} Q \, \frac{H_p}{3960} \tag{3.5}$$

The horsepower required to drive the pump is referred to as brake horsepower and is defined as

$$\text{bhp} = \frac{\text{whp}}{e_p} \tag{3.6}$$

in which $e_p$ is the efficiency of the pump ($e_p = \text{whp}/\text{bhp}$).

The total input horsepower to the motor is

$$\text{hp} = \frac{\text{whp}}{e_p e_m} \tag{3.7}$$

in which $e_m$ is the efficiency of the motor.

Electrical power consumption rate is expressed in kilowatts (kW) and is related to horsepower by $kW = hp \cdot 0.746 \, kW/hp$. Total power consumption, which is the method used to compute power charges, is expressed in kilowatt-hours (kW-h).

A graphical representation of the variation of brake horsepower with flow rate for a low specific speed centrifugal pump is shown in Fig. 3.3. For a low specific speed pump, the brake horsepower typically increases with discharge. For high specific speed pumps, the horsepower near the shutoff head increases rapidly.

To determine the water horsepower and select the best operating point for the pump, it is necessary to specify the efficiency of the pump (also shown in Fig. 3.3). From data such as shown in Fig. 3.3, it is possible to predict the input and output horsepower for any discharge, determine the bep, and decide if the pump is stable. The point of maximum efficiency is also called the design point and identifies $H_{\text{des}}$ and $Q_{\text{des}}$. When selecting a pump, it is desirable to have it operate near its design point or bep.

When selecting a pump for a particular application, it is usually possible to select from several impeller diameters, such as curves A, B, and C in Fig. 3. Curve A is for the largest impeller diameter. The option of different impeller diameters allows more flexibility in choosing a pump that meets the system requirements and operates near its design point. Each impeller has a separate bhp line, but usually a common NPSHr (net positive suction head required) line.

**Net positive suction head (NPSH)**

Satisfactory pump performance requires that adequate attention be given to cavitation. Pumps can be forced to cavitate by reducing the suction pressure. Cavitation has two general effects on pump performance. First, the cavitation can cause erosion damage, which wears away the impeller and other parts of the pump and eventually degrades the pump performance. Second, for advanced stages of cavitation, even before erosion has had time to occur, the pump performance can be degraded by large quantities of vapor.

The pressure necessary at the suction side of the pump to prevent cavitation from deteriorating the pump performance is referred to as the net positive suction head required (NPSHr). The NPSHr is determined from pump tests. It is essential that the net positive suction head available (NPSHa), which depends upon the system, exceeds the required NPSHr with a reasonable margin of safety to ensure satisfactory operation. This margin of safety is discussed in section 3.6.

For a pump connected to a suction reservoir, the NPSHa is calculated from

$$\text{NPSHa} = H_b - H_{va} + z_s - H_f \qquad (3.8)$$

in which $H_b$ is the minimum expected absolute barometric pressure head, $z_s$ the elevation from the centerline of the pump suction to the water surface elevation in the suction well (negative if the water surface is below the pump), $H_f$ the friction head loss and any local losses in the suction piping, and $H_{va}$ the absolute vapor pressure of the liquid at the maximum expected water temperature. All units in Eq. 3.8 are expressed in feet (or m) of fluid. In the case of pumps where there is no suction well, and therefore no water surface elevation for reference, the quantity $z_s - H_f$ is replaced by the gauge pressure head $P_s/\gamma$ at the centerline of the pump suction plus the suction velocity head, so

$$\text{NPSHa} = H_b - H_{va} + \frac{P_s}{\gamma} + \frac{V_s^2}{2g} \qquad (3.9)$$

The definition of NPSHa and the relationship between the two equations (3.8 and 3.9) is illustrated in Fig. 3.4. Note that the NPSHa is the vertical distance between the absolute vapor pressure line and the energy grade line (EGL).

To demonstrate the application of NPSH, consider the following example.

**Example 3.1.** Consider a pump installed so it is above the water level in the supply reservoir. The problem is to find the maximum elevation ($z_s$) of the pump so NPSHr = NPSHa.

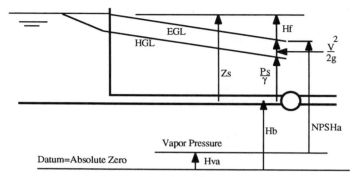

**Fig. 3.4**   Graphic description of NPSHa.

The problem is solved by determining the four quantities on the right side of Eq. 3.8. Assume $Q = 8$ cfs, maximum water temperature = 40°F, site elevation = 5000 ft, suction diameter = 12-in., loss coefficients are $K_l$ entrance = 0.9 and $K_l$ elbow = 0.7, $fL/d = 0.3$, and NPSHr = 20.5 ft.

Using Eq. 1.11 for an elevation of 5000 ft at the pump, the average barometric pressure is 12.2 psi or 28.2 ft.

Head loss in the suction pipe calculated with Eq. 3.1 gives: $H_f = 3.07$ ft.

The absolute vapor pressure at the maximum expected water temperature of 40°F is obtained from Table 1.1 and equals 0.28 ft. By setting NPSHa = NPSHr = 20.5, the solution of Eq. 3.8 gives $20.5 = 28.2 - 0.28 + z_s - 3.07$ or $z_s = -4.35$ ft, which means that the water level can be 4.35 ft below the pump. It would therefore be recommended that the pump be placed several feet lower to reduce the cavitation.

Another factor not included in this example which is necessary for determining the pump setting is water level fluctuations in the sump. The NPSHa should be established relative to the lowest possible water level in the sump. It should also be noted that NPSHr usually increases with flow rate (Fig. 3.3), so the NPSHr scaled from the pump curve should be based on the maximum expected pump discharge.

The potential of a pump to cavitate can also be expressed as a dimensionless cavitation parameter $\sigma$:

$$\sigma = \text{NPSHr}/H_p \qquad (3.10)$$

The method of evaluating NPSHr or $\sigma$ is discussed in Section 3.6. It represents the condition at which the cavitation is severe enough to cause the pump efficiency to decrease. This condition can only be evaluated by testing the pump. It is important to realize that at this condition the pump will usually suffer some cavitation erosion damage.

### Specific Speed

Specific speed is a parameter which correlates pump capacity, head, and speed as follows:

$$N_s = N \, \frac{Q^{0.5}}{H_p^{0.75}} \tag{3.11}$$

in which $N_s$ is the specific speed, $N$ the rotational speed of the pump in revolutions per minute (rpm), $Q$ the flow rate in U.S. gallons per minute (gpm), and $H_p$ is the total dynamic head in feet. $Q$ and $H_p$ are for optimum efficiency (bep).

Pumps are divided into three general classes depending on the nature of the flow pattern inside the pump and the magnitude of the specific speed. Radial flow or turbine pumps produce large heads and small discharges and have small values of $N_s$ (500–2,000). Mixed-flow pumps produce modest head increases and reasonably large flows and have intermediate values of $N_s$ (2,000–7,000). Pumps that produce large amounts of discharge at relatively low head have large values of $N_s$ (7,000–15,000) and are referred to as axial flow or propeller pumps. For a given pump design, the specific speed can be altered by changing the pump speed. Typical values of pump speed are 450, 900, 1,800, and 3,600 rpm. The high speeds are associated with smaller pumps.

When selecting the speed, two opposing factors must be considered. A larger $N$ produces a larger $N_s$ and for $N_s < 2,000$ improves the efficiency. The higher speed also results in a smaller pump and less cost. The disadvantages to higher speed are faster wear (especially if there are suspended solids in the water) and increased problems with cavitation and transients. The wear is approximately proportional to the square of the shaft speed, so doubling $N$ may increase wear by four times, which increases maintenance costs.

### Suction Specific Speed

This is another parameter used to describe the cavitation characteristics of an impeller. It is defined as

$$S = \frac{NQ^{0.5}}{\text{NPSHr}^{0.75}} \tag{3.12}$$

This parameter relates the cavitation potential of a pump to its speed and discharge. $N$ is the motor speed in rpm, $Q$ the flow in gpm at the point of maximum efficiency, and NPSHr the required NPSH in feet at the point of maximum efficiency. Large values of $S$ indicate more severe cavitation conditions. A typical upper limit of $S$ for centrifugal pumps with good cavitation performance is 9,000 (30). Many commercial pumps have values of $S$ between 5,000 and 7,000. Boiler-feed and condensate pumps have values between 12,000 and 18,000 (30). Upper limits of the suction specific speed are also given by the Hydraulic Institute standards (25). It is best to obtain values of S or NPSHr for each specific pump from the pump manufacturer if they are available.

## Rotational Inertia

The quantity $WR^2$ is a parameter describing the moment of inertia of the rotating parts of the pump and motor about the axis of rotation. $R$ is the radius of gyration in feet and $W$ is the weight of the rotating parts (including the water inside the pump) in pounds. The $WR^2$ of the pump is calculated or determined experimentally by the equipment manufacturers. The parameter is used to calculate the required starting torque of the motor and for determining its coast-down speed when the power is turned off to the motor. The latter is used for hydraulic transient analysis to determine the severity of waterhammer generated when the pump is shut off.

## Similarity Laws

Scaling head and discharge data from one pump to a geometrically similar pump of a different size, or predicting performance at a different speed can be done with the following similarity equations:

$$\frac{Q}{ND^3} = \text{const} \tag{3.13}$$

$$\frac{H_p}{N^2D^2} = \text{const} \tag{3.14}$$

These equations are easily derived by dimensional analysis. They neglect viscosity but ensure similarity of the velocity vector diagrams at the impeller.

To demonstrate application of these equations, assume that it is desired to perform model tests of a large pump to evaluate its characteristics. The prototype design conditions are represented by $D_1$, $N_1$, $Q_1$, and $H_{p1}$. For the model, it is necessary to set two of the variables and determine the other two from Eqs. 3.13 and 3.14. The process may involve iteration if $Q_2$ and $H_{p2}$ are selected as the independent variables and $D_2$ and $N_2$ determined from the equations. The reason is that normally only synchronous motor speeds are used. If $D_2$ and $N_2$ are selected as the independent variables, then $Q_2$ and $H_{p2}$ are calculated directly by

$$Q_2 = Q_1 \left(\frac{N_2}{N_1}\right)\left(\frac{D_2}{D_1}\right)^3 \text{ and } H_{p2} = H_{p1}\left(\frac{(N_2 D_2)}{(N_1 D_1)}\right)^2 \qquad (3.15)$$

The equations are also useful for evaluating the influence of changing speed or impeller diameter on $Q$, $H_p$, and whp. For example, determine the effect of doubling pump speed on $Q$, $H_p$ and whp. With $D_2 = D_1$, Eq. 3.15 gives $Q_2 = 2Q_1$, and $H_{p2} = 4H_{p1}$. This causes the horsepower (Eq. 3.4) to increase 8 times.

Next, consider the case of increasing the impeller diameter 25%. From Eq. 3.15 with $N_1 = N_2$, $Q_2 = 1.25^3 \, Q_1$, $H_{p2} = 1.25^2 \, H_{p1}$ and the water horsepower increases by $(1.25)^5$.

## 3.2  PUMP SELECTION

Pumps are selected to match system requirements. Systems normally operate over a range of flow conditions due to varying demand, changes in reservoir elevations, or changes in friction or minor losses. For flexibility and reliability, it is common to use multiple pumps in parallel. For high-pressure applications, series pumps are sometimes required. Additional pumps may also be added either in series or parallel as demand increases. This section discusses single and multiple pump selection. Several examples are presented to demonstrate various principles.

When selecting a pump, the designer has many choices from different manufacturers. One selects a pump whose design point is close to the operating point and that can operate efficiently and economically over the required operating range. Consider the following example.

*Example 3.2.* Using the pump characteristic curves in Fig. 3.3, select the best impeller diameter for the folowing system. It is desired to supply 500 gpm in a system with a static lift (elevation change) of 80 ft. The pipe is 6 in. in diameter, 700 ft long, with a Darcy–Weisbach friction factor of 0.020.

The first step is to evaluate the $H$ vs. $Q$ equation (Eq. 3.1) for the system and plot it on the pump curve.

$$\Delta z = 80 \text{ ft. (given)}$$

$$H_p = 80 + CQ^2, \ C = \frac{fL}{(2gdA^2)} = 11.28$$

so

$$H_p = 80 + 11.28Q^2 \ (Q \text{ in cfs})$$

or:

$$H_p = 80 + 5.60 \times 10^{-5}Q^2 \ (Q \text{ in gpm})$$

This equation is plotted in Fig. 3.5, which contains the same pump characteristics as Fig. 3.3. The intersection of the system curve with the pump curves indicates the operating points for each impeller diameter. In terms of efficiency, either A or B impellers would work. The A impeller will supply about 550 gpm, and the B impeller will supply about 450 gpm. Since neither is exactlly 500 gpm, a decision is required. If 450 gpm is adequate, the smaller B impeller would be the economical choice. If 500 gpm is the minimum required, then the A impelled would be chosen.

For this example, pump B is selected. The efficiency is about 84.5%. The NPSH and brake horsepower are scaled by projecting vertically downward to intersect the NPSH and bhp curves, giving NPSH = 10.5 ft and bhp = 13 hp.

In most pipe systems, the elevation head varies or a valve may be throttled to change the system losses and consequently the system curve. If the elevation head changes, a new system curve is drawn by translating the original system curve vertically the amount that $\Delta z$ varies. If a valve is throttled, the value of $C$ in Eq. 3.16 must be recalculated and a new system curve plotted. The pump chosen in this example could operate at flows between about 380 and 650 gpm efficiently (above 80%).

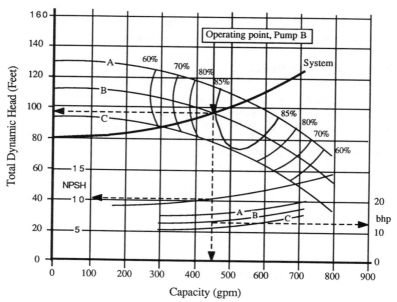

**Fig. 3.5**  Pump rating curve and system curve for Example 3.2.

Centrifugal pumps can only operate over a limited flow range on either side of the design point. If the flow gets too low, recirculation can occur at the inlet to the impeller, resulting in rough operation as well as poor efficiency. At large flows, the pump demands excessive power, which may overload the motor. Also, cavitation becomes more likely, since the NPSHr increases with $Q$, and efficiency drops. If the range of system requirements is too great, multiple pumps in parallel or series may be necessary.

**Example 3.3.** Select a pump that will supply about 6 cfs from a low reservoir (el. 1500 ft) to an upper reservoir whose water level varies between 1615 and 1640 ft. The pipe is 6000 ft long, $f = 0.016$, $d = 16$ in. (inside pipe diameter).

The required $H_p$ to satisfy the system (from Eq. 3.1) is

$$H_p = \Delta z + 0.573 Q^2, \ (Q \text{ in cfs})$$

where $\Delta z = 140$ to 115 ft, depending on the reservoir.

Consider a pump having $H-Q$ characteristics defined by the equation

$$H_p = 175 - 0.5Q - 0.4Q^2 \ (Q \text{ in cfs})$$

Find the discharge at maximum $\Delta z = 140$ ft. This is done by simultaneously solving the above equations, which is analogous to finding the intersection of the pump and system head curves in Fig. 3.5.

$$140 + 0.573 Q^2 = 175 - 0.5Q - 0.4Q^2$$

The real root is $Q = 5.75$ cfs.

At minimum $\Delta z = 115$ ft. the flow rate (using the same approach) increases to 7.49 cfs. The pump will therefore supply between 5.75 and 7.49 cfs depending on the water surface elevation in the terminal reservoir.

To complete this analysis, one should check the pump curve to be sure that the range of flows fall in the zone of acceptable efficiency and check the NPSHa for the system to be sure that the cavitation will not be excessive.

## Parallel Pump Selection

There are a number of reasons for placing multiple pumps in parallel. For reliability, it is desirable to have more than one pump. It is common to use three identical pumps in parallel, each having the capability of supplying

50% of the normal flow requirement. Another option is to have four pumps, each capable of supplying 33% of the normal flow requirement. Each option provides a wider range of flows than a single pump, as well as a back-up pump for increased reliability. Another reason for parallel pumps is that they can be added to meet future demands. A pipeline is often designed for future demands and pumps added as the demand increases.

When pumps are operated in parallel, they work against a common pressure and it is important to match the head characteristics of pumps carefully. If the pumps are badly mismatched in head, one of them may not even produce any flow. The system head loss characteristics are also important because they help determine the type of pump characteristic curve that is most suitable. It is also important that the pumps are able to operate efficiently individually or together. These principles will be discussed in the following examples.

Figure 3.6 demonstrates the situation of two identical pumps in parallel. The combined pump curve is constructed by adding (doubling if the two pumps are identical) the discharge at each head. The design head or bep of the pump occurs at the same head for one or two pumps (if the pumps are identical). The pump should be selected so that at the most frequent system operating condition the pump(s) will be operating as closely as possible to their design point for both single- and multiple-pump operation. Pump selection is further complicated because the static lift can vary anywhere between the maximum and minimum, so there is usually a family of system head curves. The pump characteristics used in Fig. 3.6 are the same as in Fig. 3.3.

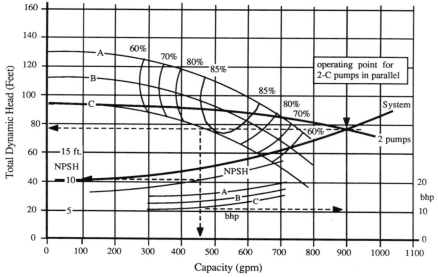

**Fig. 3.6**  Parallel pump system head curves for identical pumps.

*Example 3.4.* For the system curve shown in Fig. 3.6, find the flow rate for two type C pumps operating in parallel.

The characteristic curve for two identical pumps in parallel is constructed by doubling the discharge for a single pump at any head. The actual discharge may be slightly less due to minor losses associated with the more complex piping required for multiple pumps. This is adjusted for by increasing the value of $C$ in the system loss equation (Eq. 3.1b).

The intersection of the two pump curve and the system curve is the operating point. The flow for two pumps will be 900 gpm at a head of 78 ft. The efficiency, NPSH, and horsepower for each pump is found by projecting back to the single-pump curve along a line of constant head, as shown by the broken line in Fig. 3.6. Reading data from the single pump curve gives

Efficiency = 84%, NPSHr = 10.5  ft, and bhp = 10.5 hp for each pump

If one pump were operating in the same system, its flow and head would be determined at the intersection of the single pump curve for pump C and the system curve. Figure 3.6 shows that the head and flow would be 59 ft, and gpm 640 and efficiency 68%.

If a single pump had to operate for long periods of time, it could operate at a more efficient point by throttling a valve to raise the system curve. For example, a flow of 500 gpm, which would put the pump at its best efficiency point (bep), could be obtained by creating a head loss across a valve of about 25 ft.

The next example demonstrates the problems associated with using pumps in parallel that have different head–discharge curves. Normally, pumps operated in parallel should either be identical or have very similar characteristic curves.

*Example 3.4, continued.* Using the same system and pump curves just discussed, consider using one type-B pump and one type-C pump in parallel.

The combined pump curve, shown in Fig. 3.7, is again obtained by adding the flows at constant head. However, at heads below about 93 ft, pump C cannot supply water because the pressure is above its shutoff head. The combined pump curve merely follows the B pump curve until the head drops below 93 ft. Beyond that point, the discharge from the two pumps add, as shown in Fig. 3.7.

In terms of operation at heads above 93 ft, if pump C were turned on and there were no check valve to prevent reverse flow, water would

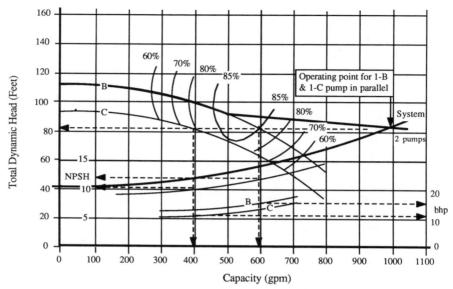

**Fig. 3.7** Parallel pump system head curves for mismatched pumps.

flow backward through pump C even though it would be trying to pump forward. With a check valve in the discharge pipe, the valve would close and pump C would pump against a closed valve. This would rapidly overheat the pump and cause damage.

One could handle this with controls that prevent pump C from operating at heads above some set value. The better solution is to match the heads so this situation does not happen.

For the mismatched pumps in parallel, the flow is increased to 990 gpm. The efficiency, NPSHr, and bhp are scaled from each individual pump curve, as shown in Fig 3.7, giving

For pump B: $Q = 590$ gpm, $H_p = 81$ ft, efficiency $= 85\%$, NPSHr

$= 12$, and bhp $= 15$ hp

For pump C: $Q = 400$ gpm, $H_p = 81$ ft, efficiency $= 80\%$, NPSHr

$= 10$, and bhp $= 12$ hp

### Series Pumps

There are several situations in which pumps in series may be useful. Consider the data shown in Fig. 3.8. It represents a system with a high,

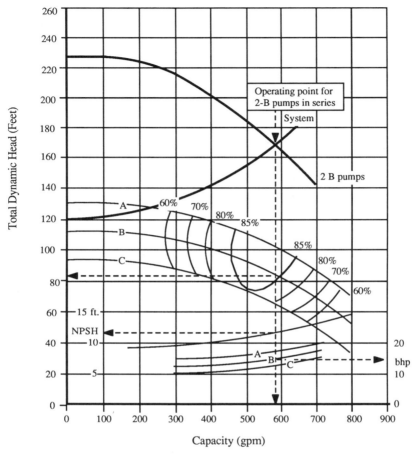

**Fig. 3.8**   Pump rating curve for series pumps for Example 3.5.

constant static lift of 120 ft and high friction losses. Figure 3.8 again shows the same three impeller sizes used in the previous examples. It is apparent that none of the pumps will supply adequate head. An obvious solution is, of course, to select another type of pump that will have adequate head. For this example, two pumps in series will be used to demonstrate the principle. The system curve for two identical pumps in series is created by doubling the head for one pump at each flow. Figure 3.8 shows the combined pump curve for two B pumps intersects the system curve at a flow of 580 gpm and a head of 168 ft.

One must project back to the single-pump curve to read efficiency, brake horsepower, and NPSH. For parallel pumps, this was done by projecting to the left along a line of $H_p$ = constant. For series pumps, this is done by projecting down at $Q$ = constant. Once on the single-pump line, efficiency

$(e_p)$, NPSHr and bhp can be determined. The pumps would be operating near their bep at an efficiency of 85% at $H_p = 82$ ft (each), $Q = 580$ gpm, NPSHr $= 12.3$ ft, and bhp $= 15$ hp (each). For this system, it would not be possible to operate one pump alone because the elevation lift of 120 ft exceeds the shutoff head of the B pump.

The next example demonstrates how to use the pump characteristic equation (Eq. 3.3) and the system equation (Eq. 3.1) to solve for parallel and series pump flows mathematically, as opposed to graphically as done in the previous examples.

**Example 3.5.** Consider a pipeline connecting two reservoirs. Assume that the system characteristics are expressed by $H_p = 60 + 1.2Q^2$, in which $H_p$ is in feet, 60 is the elevation difference in feet between the reservoirs, and the friction loss is $1.2Q^2$ ($Q$ in cfs). The pump selected for the application has a $H$–$Q$ characteristic expressed as $H_p = 80 - 0.8Q - 0.6Q^2$ ($Q$ in cfs). Find the flow rate for a) one pump, b) two pumps in series, and c) two pumps in parallel.

For one pump, the flow rate produced by the pump is determined by simultaneously solving the system and pump H-Q equations

$$H_p = 80 - 0.8Q - 0.6Q^2 = 60 + 1.2Q^2$$

The real root is $Q = 3.11$ cfs and the corresponding head is 71.6 ft.

For two pumps in series, the head at any flow is approximately double, so:

$$H_p = 160 - 1.6Q - 1.2Q^2 = 60 + 1.2Q^2$$

The real root is $Q = 6.13$ cfs and the corresponding head is 105 ft.

For two pumps in parallel, the discharge is half through each pump at the same head, so

$$H_p = 80 - 0.8(Q/2) - 0.6(Q/2)^2$$

The real root is $Q = 3.70$ cfs and the corresponding head is 76.4 ft.

Using this procedure, it is still necessary to refer to the pump curves to see if the pumps are operating in an efficient range and to get efficiency, bhp, and NPSHr.

The shape of the system curve has an important bearing on pump selection for single or multiple pumps in series or parallel. Figures 3.9–3.12 demonstrate several features of the influence of system characteristics on pump selection. In these figures, the bep has been arbitrarily set midway between the operating points for one- and two-pump operation.

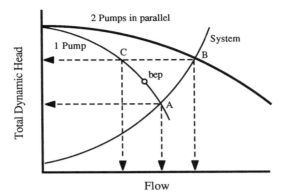

**Fig. 3.9**  Parallel pumps in high friction loss system.

Figure 3.9 shows the operation of two pumps in parallel in a system that has high friction losses and little static lift that is, a steep system curve. Point A is the operating point for a single pump. Point B identifies the operating point for both pumps operating. Point C is the projected point for each pump with both pumps operating. It is located by projecting to the left at constant head.

Note that the flow is only increased about 30% by using two pumps. It is assumed that the bep lies half way between points A and C, as shown in the figure. Since neither point A nor C are very close to the bep and the flow is only increased 30%, using parallel pumps in high friction systems is not very efficient.

Next, consider the same two pumps operating in a system with a large gravity lift and low friction, as shown in Fig. 3.10. For this case, the flow is almost doubled with two pumps in parallel. Also note that points A

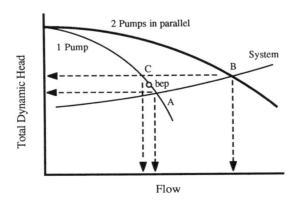

**Fig. 3.10**  Parallel pumps in low friction loss system.

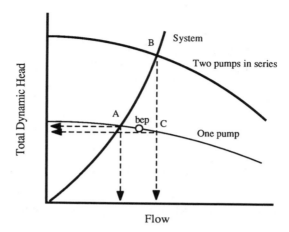

**Fig. 3.11**   Series pumps in high friction loss system.

(one-pump operation) and C (two-pump operation) are both near the bep. This makes for more flexibility, since either one or two pumps can be operated and still have good efficiency. These two examples demonstrate that parallel pumps are more appropriate in systems with low friction loss.

The next two figures demonstrate that series pumps are better for high friction loss systems. Figure 3.11 shows series pumps in a high friction system, while Fig. 3.12 shows series pumps in a low friction system. Point B defines the operating point for two-pump operation. Point C identifies the head and flow for each of the pumps operating together. Point A is for one pump operation. For the high loss system (Fig. 3.11), the flow is only

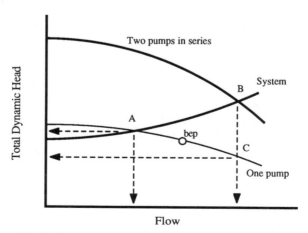

**Fig. 3.12**   Series pumps in low friction loss system.

increased about 50% when operating two pumps, but points A and C are both near the bep. Therefore, either one or two pumps can be used efficiently. This would be a good application for series pumps.

When series pumps are used in a low friction system, as in Fig. 3.12, the flow is almost doubled, but the pumps operate far from the bep for both one- and two-pump operation. This would not be a good application for series pumps. Series pumps work better in high loss systems and parallel pumps better in low-loss systems.

## 3.3   PUMP OPERATION

There are mechanical, electrical, and hydraulic problems associated with starting and stopping pumps. If a pump is started against a closed valve, as soon as it is up to speed the pressure goes to the shutoff head. The piping between the pump and valve must be designed for this pressure. If a low specific speed pump is allowed to operate against a closed valve for any period of time, the energy transferred to the water can cause it to boil and possibly damage the pump.

The temperature rise of the water inside the pump case caused by operating a pump at shutoff head, ignoring heat loss by radiation and convection, can be determined from the equation (Ref. 30, pp. 13–12):

$$T_{rm} = (42.4 P_{so})/(W_w C_w) \qquad (3.10)$$

where

$T_{rm}$ = temperature rise, °F/min

$P_{so}$ = brake horsepower at shutoff

42.4 = conversion from bhp to Btu/min

$W_w$ = net weight of liquid in the pump, lb

$C_w$ = specific heat of liquid (1.0 if liquid is water)

For high head pumps, this temperature rise can endanger the pump in minutes.

High specific speed pumps should not be operated at shutoff head against a closed valve for any period of time because the power and torque are considerably higher than the rated values. Since they require excessive power and torque, the motor will overheat rapidly. This problem can often be solved by installing a check valve which opens as soon as the head exceeds the static shutoff head.

For low specific speed pumps (below about 5000), the power and torque at shutoff are equal to or less than the rated conditions. Starting a low specific speed pump against a closed valve will not overload the motor. It is still necessary to limit the time of operation at shutoff so the water does not heat up.

If starting with a closed valve is necessary to control the filling rate, it is often advisable to have a bypass line that returns part of the flow to the suction sump or to the suction piping. The bypass would be designed to provide enough recirculating flow to reduce the head and power to acceptable limits. The bypass line must be connected far enough from the pump suction that it does not disturb the flow into the impeller. Bypass valves can be manually operated or hydraulically actuated by line pressure.

For a system with a check valve, water starts to flow as soon as the pump head exceeds the gravity head against the downstream side of the valve. The acceleration of the water can be calculated by the equation of motion, $\Sigma F = ma$. The forces resisting motion are gravity and friction. The force causing motion is the pump pressure times the pipe area. The mass is the total mass of water in the pump and pipe that must be accelerated. It is a simple procedure to evaluate the time required to reach steady state. The time depends on the specific speed and the amount of gravity head. For lower specific speed pumps, the shutoff head is not much larger than the design head. If there is a large gravity head, the net head available to accelerate the water is small and the time to reach steady state large. For a high specific speed pump, the shutoff head is several times the design head. Such pumps usually pump primarily against friction. At low flow rates, there is little friction and consequently little resisting force, and the accelerating head is high. This system will reach steady state rapidly.

Because starting pumps causes electrical surges, most power companies require reduced voltage starters on all but small pumps. The cost of such starters can be a significant percentage of the total cost and must not be ignored in the economic analysis.

Hydraulic transients are always generated in pump systems at start-up and shutdown. If special precautions are not taken, the magnitude of the transient can be sufficient to cause damage. For example, consider a system where the pump is started with the discharge pipe empty. If there is not a control valve to regulate the flow through the pump at start-up, the pump will initially supply more than the design discharge, since there is virtually no head on the system. Assume that a control valve at the end of the pipe is only partially open. When the pump starts, water will rapidly fill the pipe and the air will be expelled through the valve. Air can easily pass through the valve with little resistance. When the water reaches the valve, because it is much more dense than the air, it cannot pass through the valve as easily and therefore there is an almost instantaneous decrease in velocity which can create a high pressure rise.

Start-up transients can be controlled by starting the pump against a closed or partially open discharge valve located near the pump and using a bypass line around the pump. This allows the discharge in the system to be filled slowly. For systems where the discharge piping remains full because of a

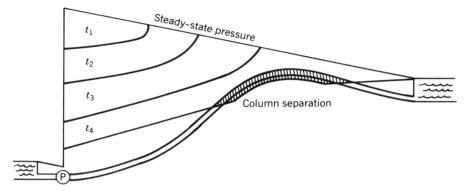

**Fig. 3.13**   Transient caused by power failure to a pump.

check valve, it is usually safe to start the pump without using a throttled discharge valve.

The problem of shutting the pump down also creates hydraulic problems. When the power to a pump is turned off, the head at the pump rapidly decreases and flow reversal can occur. The magnitude of the transient is related to the length and profile of the discharge line, the pump characteristics, and the relationships between the elevation and friction head.

One transient problem is related to systems pumping over a hill. Such a system is shown in Fig. 3.13. For steady-state flow, the pressure along the pipe is shown by the straight sloping line. When the pump is turned off, the pressure at the discharge of the pump rapidly decreases and this sends negative pressure waves down the pipeline. At the top of the hill, the pressure is relatively low. When the negative pressure wave reaches the top of the hill, the pressure in the pipe can drop to vapor pressure. This is referred to as column separation. A vapor cavities forms near the top of the hill. On subsequent cycles of the transient when the pressure increases, cavity collapse and may occur and cause a high pressure to be generated in the system.

Numerous mechanical devices and techniques can be used to suppress pump shutdown transients. These include increasing the $WR^2$ of the pump, use of air chambers near the pump, pressure-relief valves, vacuum-breaking valves, surge tanks, and sophisticated interaction between control valves and the pump. Methods of computing transients and devices for controlling them are discussed in later chapters.

## 3.4   SUCTION PIPING

Flow conditions at the inlet to the pump can significantly affect pump performance. Good flow conditions in suction piping require adequate suction pressure (so that the available NPSH is high enough) and reasonably steady and uniform flow at the impeller. Elbows, tees, and valves in the suction piping cause significant disturbance to the flow. If close to the pump, they cause the flow to be rotational, nonuniform, and highly turbulent. These effects reduce pump performance and cause cavitation. Attention must be given to providing a satisfactory velocity distribution with a low level of turbulence at the pump suction.

The other concern with suction piping is maintaining adequate pressure. This is done by using a larger diameter for the suction pipe to reduce losses, placing the pump at the lowest possible elevation, or by placing a booster pump upstream.

## 3.5  HYDRAULICS OF PUMPING PITS

The Hydraulic Institute publishes standards for sizing pump intakes for vertical pumps (25). The standard recommends general dimensions and submergence depth. For small pumps, following these standards have generally produced a satisfactory installation. As the size of the units increases, so do the hydraulic problems.

The general design criteria require that the flow approaching the pump be uniform and steady. If the pit is too small, disturbances caused by the flow entering the pit can cause rough pump operation. If the inlet to the pit is too close to the pump, the approach flow will be nonuniform and highly turbulent, again causing rough pump operation. When multiple pumps are placed in the same pit, they should be located so that the flow does not pass by one pump to get to another. If the inflow free falls into the pit, care should be given to prevent air from being entrained in the liquid. If the liquid contains more than about 3% free air, the pump will lose efficiency and operate roughly. The submergence must be adequate to prevent surface vortices. The pumps must not operate at flows too far below or above the rated flow.

Model studies of large wet-pit-type pump installations will normally identify some or all of the following problems that result in poor pump performance (68):

1. Aerated and nonaerated surface vortices.
2. Subsurface vortices from the floor and side walls.
3. Circulation in the pumping pit caused by a nonuniform approach flow that generates vortices and prerotation at the throat of the pump.
4. Flow separation from piers, entrances, or any discontinuity of the boundaries that generates turbulent eddies.
5. Flow separation at the suction bell, lower bearing, and support struts.
6. Inadequate water depth for suppressing cavitation.
7. Too wide an operating range for the pump so that it operates part time at flows much different than its rated flow.

Surface vortices are one of the most widely recognized hydraulic problem encountered in large wet-pit-type pump installations. They are easily detected in both the model and prototype and have an immediate and easily recognizable influence on pump performance. They can cause vibrations, reduced efficiency of the pump, and additional problems in the discharge piping. Depending on the geometry of the pumping pit, capacity of the pump, and water surface elevation, these vortices can be either aerated or nonaerated. Their stability and strength depend on numerous factors that are difficult to predict without a physical model. The aerated surface vortex would likely cause little cavitation, since it is aerated, but can reduce efficiency and generate vibrations due to the variation in the density and velocity as the impeller cuts through the vortex.

Depending on the flow rate and the relative distances between the suction bell of the pump and the boundaries of the pumping pit, subsurface vortices can be found coming from the floor and the side walls. The strength and location of these vortices depend primarily on the flow rate, circulation within the pumping pit, and the relative distances between the suction bell and the boundaries.

The subsurface vortices coming from the side and rear walls generally concentrate near the suction bell intersecting the impeller blades at their extremities, where the velocity is maximum. As a result, they can cause significant cavitation and vibration damage, especially in larger units.

The floor vortex is common to almost all installations unless special provisions have been made to eliminate it. If there is any measurable circulation within the pumping pit, the floor vortex is usually strong and stable with a constant direction of rotation. If the flow in the pumping pit is generally uniform with little circulation, the floor vortex can be unstable, changing direction frequently. It is not uncommon to find two floor vortices somewhat equal in strength and opposite in rotation. The floor vortex is usually concentrated near the centerline of the pump so that it intersects the impeller blades near the hub. Cavitation erosion damage near the impeller blades on the hubs of certain pump installations could be a result of the floor vortex, especially if damage is observed on both sides of the blades.

Circulation has two detrimental effects on pump operation. First, it has a significant influence on the number, size, and location of vortices generated within the pumping pit. Second, it causes prerotation at the throat of the pump, which changes the angle of approach of the flow hitting the impeller blades. Since impeller blades are designed on the assumption that there is one dimensional axial flow at the throat of the suction bell, prerotation may cause separation and possible cavitation at the leading edges of the impeller blades.

When designing large capacity pump installations, the preceding problems should be anticipated early in the design process. The advantages of a model study should be seriously considered (61).

If the intake is on a river or lake, sediment, trash, and fish exclusion will need to be considered. There are several approaches to the fish problem. One is to design so that fish can swim near the intake and not be trapped against the screens. This generally results in large intakes with high cost. Acoustical noises have been used in an attempt to scare the fish away from the intake. A recent approach has been to trap the fish and move them safely away from the intake. For details on the current state of the art as of 1984, the reader is referred to Reference 2, a publication by a task committee of the American Society of Civil Engineers on fish handling. In cold regions, icing at the intake is also a design consideration.

## 3.6   CAVITATION IN PUMPS

One of the serious problems with pump operation continues to be cavitation erosion. Such damage causes financial loss due to downtime and repair or replacement costs. Most pumps operate in cavitation. The goal of pump selection and proper installation is to limit the cavitation damage to an acceptable amount. Assuming that the pump is properly designed, the cavitation is usually generated due either to poor suction conditions, insufficient NPSHa, or operating the pump at conditions incompatible with its rated design point. A major factor contributing to cavitation erosion damage is that it is standard in the industry to operate pumps just slightly above their required NPSH value. Since NPSHr is the point at which efficiency of the pump drops off due to heavy cavitation, operating the pump near that condition can result in cavitation erosion damage.

Cavitation, whether occurring in pumps, valves, or orifice plates, generates noise, pressure fluctuations, vibrations, and eventually reduces the efficiency of the device. Identifying the early stages of cavitation in a pump can be done with pressure transducers, hydrophones, accelerometers, or visual inspection using a stroboscope. It is helpful to use a high-frequency pickup and filter out the low frequency disturbances generated by the pump.

Reference 16 reports a study in which flush-mounted hydrophones were used to detect the onset of cavitation in pumps. By analyzing the output above 8 kHz, incipient cavitation point was identified which corresponded closely to a visual onset of cavitation. Below 8 kHz, the interference from mechanical and flow noise masked the cavitation noise and made evaluation of an incipient cavitation point difficult.

To evaluate NPSHr, a test pump is set at a given discharge with a high positive suction pressure. The head and horsepower are then measured. In successive steps, the suction pressure is lowered while the discharge is maintained constant and the total dynamic head measured for each condition. The head remains constant until the cavitation becomes heavy. As the suction pressure is decreased, cavitation is first detected either visually or by ear. As the suction pressure is continually lowered, the noise, vibrations, and cavitation erosion potential continue to increase. Once the cavitation is severe enough that fairly large vapor pockets form on the impeller, the head and efficiency begin to reduce and the noise and vibrations become lower in frequency. Once this condition is reached, only slight changes in the suction control valve cause the suction pressure to drop to vapor pressure and cause the efficiency of the pump to be significantly reduced. The net positive suction head required (NPSHr) is usually picked at the point where $H_p$ has dropped 3% from its normal value. Typically, the net positive suction head is a minimum at lower flow rates and increases as the discharge increases.

The relationship between NPSHr, cavitation damage, and noise generated by cavitation has been investigated (16, 21). Reference 16 shows that maximum noise and erosion rates occur before the efficiency of the pump begins to decrease. Reference 21 suggests that on the average, NPSHa should be about three times NPSHr to prevent any cavitation. These findings are also substantiated in Reference 16. The data correlate with the author's observations of tests on numerous valves and orifice plates: maximum cavitation erosion, noise, and vibration occur before the device begins to choke.

Tests (16) were conducted on nine different pumps to measure onset of cavitation and NPSHr (called the breakdown point, or loss of efficiency point). The data showed that ratio of NPSH at inception of cavitation to NPSH at breakdown (NPSHr) varies from 2.2 to 18.3. Onset of cavitation varied significantly with flow rate. For one pump at design flow, the NPSH ratio was 8.24, but at half the design flow it increased to 18.3.

It is not recommended that pumps be operated so that NPSHa is several times NPSHr because the level of cavitation at inception is insignificant. What the data do suggest is that since cavitation usually begins long before there is a loss of efficiency, there is also a good chance that significant damage will occur at NPSH > NPSHr. The present practice of setting NPSHa just a few feet higher than NPSHr may not be safe. Hopefully, additional testing should be done to evaluate onset of cavitation damage for pumps to clarify this point.

## PROBLEMS

**3.1.** Why is the suction side of a centrifugal pump usually larger than the discharge side?

**3.2.** Explain why point 3 on Fig. 3.2, curve B, is the boundary between stable and unstable pump operation.

**3.3.** Pressure gauges on the suction and discharge sides of a pump read 5 psi and 108 psi, respectively. Can you calculate or estimate the total dynamic head?

**3.4.** An 8-in. suction by 6-in. discharge pump is being tested in the laboratory. The head rise across the pump is measured with a water/mercury differential manometer. At a flow rate of 3 cfs, the manometer reads 215-cm differential. (a) Calculate the total dynamic head (TDH). (b) What would happen to the TDH if friction losses were included in the calculation? (c) What is the water horsepower?

**3.5.** When basing the available NPSH on the measured suction pressure $(P_s)$ Eq. 3.9 is used; prove Eq. 3.9 is valid. Hint: apply the energy equation between the suction well and the pump inlet.

**3.6.** A pump is to be installed in a straight horizontal pipe. The site is at an elevation of 3000 ft, the water temperature is 50°F and the flow rate will be 5 cfs. The required NPSH is 15 ft. Calculate how far the pump can be located from the suction reservoir (el. 3010 ft) so that the available NPSH is equal to the required NPSH. The pipe diameter is 12 in. Assume $f = 0.0145$.

*Answer:*    2740 ft.

**3.7.** As the speed of a pump is increased, what happens to: cost, wear, specific speed, efficiency, and transient problems?

**3.8.** A commercial centrifugal pump is being considered for an application where $N = 1750$ rpm, $Q = 9000$ gpm, and NPSHa = 25 ft. What is the possibility of cavitation?

**3.9.** Using the data for the 24-in pipe in Example 2.5, find the system–head curve (Eq. 3.1b).

*Answer:*    $H_p = 175 + 0.107Q^2$

**3.10.** Determine the head–discharge equation (Eq. 3.3) for a pump that produced the following data during a test:

| $Q$ (cfs) | *Head* |
|:---:|:---:|
| 0 | 270 |
| 10 | 250 |
| 25 | 180 |

Using the system equation determined in Problem 3.9, find the operating point for this pump.

*Answer:*    (a) $H_p = 270 - 0.933Q - 0.1067Q^2$
(b) $Q = 19$ cfs, $H_p = 213.6$ ft

**3.11.** A system has a required head of 135 ft at 600 gpm. The elevation lift is 50 ft. Using the pump curve in Fig. 3.3, find the smallest pumps in series or parallel that will supply the required flow. List $Q$, $H_p$, efficiency and horsepower for each pump.

**3.12.** A pump with a $H - Q$ curve of $H_p = 125 - 0.4Q - 0.3Q^2$ (cfs) is installed in a system defined by $H_p = 50 + 0.41Q^2$. Find the flow.

# 4

# VALVES

Valves are an important part of pipeline design. They are used to regulate the flow and pressure, protect the pipe and pumps from overpressurization, help to prevent transients, prevent reverse flow through pumps, remove air, and perform various other functions. If not properly selected and operated, however, they can also cause problems. Closing a control valve too fast, using the wrong type of check valve, or filling a line too rapidly can result in severe hydraulic transients. If valves are subjected to cavitation, they will wear out rapidly, leak, and need replacing. This chapter discusses types and functions of valves and their hydraulic characteristics, including energy dissipation and torque. Problems of cavitation and transients are covered in later chapters.

There is a wide assortment of different types of valves used for a variety of purposes. Table 4.1 lists typical uses. For the sake of discussion, valves and their uses will be separated into four categories: 1) control valves, 2) pressure regulating valves, 3) nonreturn flow valves, and 4) air control valves. These categories are not exclusive because the same valve type may be used with different controls to perform any of the four functions. This same overlapping between categories applies to the valve uses listed in Table 4.1. For example, a flow-regulating valve should be designed so it does not produce excessive cavitation. This may require multiple valves in series so they function as both flow regulating and cavitation control valves.

## 4.1 CONTROL VALVES

The term control valve refers more to the function than the style of the valve. It can refer to any valve that serves to regulate the steady-state flow

**TABLE 4.1  Typical Uses for Valves.**

| Valve Type | Function |
|---|---|
| *Control valves* | |
| Energy dissipation flow and cavitation control | Control flow or dissipate excess energy while limiting cavitation to an acceptable level |
| Isolation and sectionalizing | Isolate a pump or control valve or isolate a section of a long pipe for repairs. |
| Free discharge | For release of water from reservoirs and bypass at turbines for transient control |
| Bypass | For recirculation at a pump start-up, filling a pipeline, an equalizing pressure across large valves or gates prior to opening |
| | |
| *Pressure regulating valves* | |
| Pressure-relief | Protects pipe from excessive pressures |
| Surge-anticipating | Similar to a pressure-relief valve but activated by a signal indicating that a transient has been generated |
| Pressure-reducing or -sustaining | For maintaining a constant upstream or downstream pressure; usually hydraulically activated by a pressure-sensing pilot system |
| | |
| *Nonreturn valves* | |
| Check valves | Prevent reverse flow |
| | |
| *Air control* | |
| Air relief–vacuum breaking | To remove or admit air to a pipe during a transient or during filling or draining |

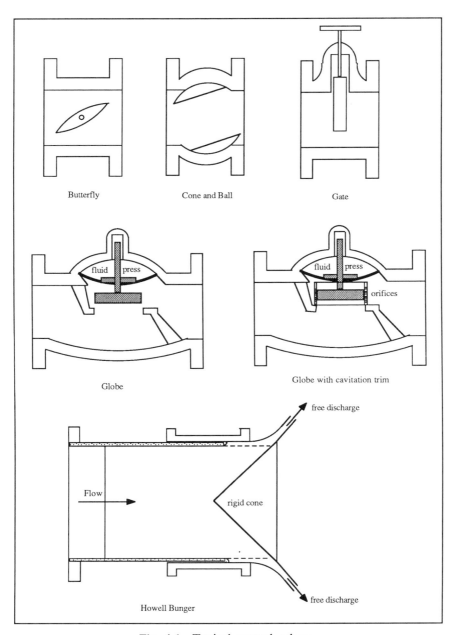

**Fig. 4.1** Typical control valves.

or pressure in a system. This includes isolation, block, or sectionalizing valves that are used to prevent flow in certain sections of the pipe. Of more interest are control valves used to control the flow without creating undesirable transients, excess cavitation, or head loss and able to function under all expected flow conditions.

There are numerous valves that can be classified as control valves. A brief description of the common types follows. A thorough treatment of many types of gates and valves is presented in Chapter 22 of Reference 14. Figure 4.1 contains simplified sketches of several types of valves.

### Gate Valves

This type of valve has a totally enclosed body with a circular or rectangular disk or gate that moves perpendicular to the flow direction. Some gate valves have a tapered circular disk and tapered guide slots. The matching taper of the valve seat causes a watertight metal to metal contact as the disk is wedged into the seating surface. The disk is normally raised by rotation of a hand wheel. The gate valve also comes in a double-disk gate design such that when the valve is closed both sides of the disk are wedged against the seats.

When a full-ported gate valve is wide open, the flow passage is only slightly less than the area of the pipe due to projecting seats and guides, so it has a large discharge capacity and small pressure drop. Some gate valves have a reduced port where the valve body is formed into a venturi shape, causing the fully open area to be reduced. Gate valves are normally operated only wide open or closed and not for regulating flow.

### Butterfly Valves

A typical butterfly valve, shown in Fig. 4.1, consists basically of a disk which rotates 90° from fully open to fully closed. There are numerous alternative disk designs. This includes symmetrical (as shown in Fig. 4.1), unsymmetrical, eccentric, and flow-through. The shape of the disk influences the capacity and flow torque.

The butterfly is a popular valve at the present time due to its light weight, compact size, satisfactory performance, and low cost. The valves are suitable for throttling as well as on–off service. With certain disk designs, the flow capacity of a butterfly valve can approach that of a gate valve in the full open position. Flow torque and cavitation can be controlled by alterations of the shape of the disk and the seat. A variety of materials can be used for the body, disk, and seat to make it suitable for use with almost any liquid.

### Cone, Ball, and Plug Valves

These three valves are similar in function, and hence only the cone valve is shown in Fig. 4.1. The moving part of the valve is generally conical in shape with a cylindrical hole through which the liquid flows. If it is a full-ported valve, the flow passage is the same diameter as the pipe at the valve inlet. When the valve is fully open, there is no blockage and therefore no head loss. At intermediate openings, there are two throttling ports in series, one at the inlet to the plug and one at the outlet of the plug. This gives the cone valve better cavitation characteristics than gate or butterfly valves (if the plug is skirted).

The valve plug can be either skirted or unskirted. A skirted valve plug can be either solid or hollow. For a skirted valve, all of the flow goes through the main flow passage. With an unskirted valve, flow can go around the plug as well as through it. The flow going around the plug has a single stage of dissipation. This increases the cavitation.

There are at least four different types of ball valves. Some have solid plugs with cylindrical holes. As the valve plug rotates in front of the valve port, a larger area is continually exposed. For a full-ported ball valve fully open, there is no restriction to the flow. Another style is the reduced port valve where the area of the plug is approximately 75% of the area at the inlet flange. Another option is a skeleton ball valve, which is composed of two pieces of intersecting pipe, one with sealing ends and one that is open. For this type of valve, flow can go around the plug as well as through the main port. The fourth type has a segmented ball so there is only one throttle port. The segments and skeleton ball values cavitate worse than those with solid balls.

## Globe Valves

This style of valve is suitable for a wide variety of applications for both manual and automatic control. Figure 4.1 shows a hydraulically operated globe-style valve. Normal flow is from left to right, but the valve can be operated in reverse. The valve position for a hydraulically operated globe-style valve is changed by adding or removing liquid from the chamber above the flexible diaphragm. This can be done manually or automatically with pilot controls. By changing the type of control, a globe valve can be adapted so that it maintains constant inlet pressure, constant outlet pressure, constant flow rate, and constant level in a reservoir, acts as a pressure-relief or surge-anticipating valve, and functions as a check valve. The pressurized liquid for operating the valve is normally supplied from the pressure inside the pipe.

Globe valves can also be mechanically actuated. This style is typical of smaller-diameter globe valves. In larger sizes, the load on the plug causes excessive force to be required to actuate the valve, so hydraulic control is preferred.

The globe valve has larger losses in the full open position than gate, cone or butterfly valves because of the complicated flow path. Two limitations of the valve style are that they have relatively large loss coefficients fully opened and they are only built in sizes generally up to 16 inches in diameter.

By installing additional trim inside the valve, its cavitation performance can be substantially improved. Figure 4.1 shows a typical trim. It consists of one or more cylinders containing many small orifice holes. The orifices dissipate energy and suppress cavitation. With multiple concentric cylinders, the energy can be dissipated in stages and the caviation further reduced. The disadvantage of the trim is that the loss of the valve in the full open position is significantly increased.

### Cavitation Control Valves

There are numerous types of valves that are classified as cavitation control valves. There are only two principles used in valves to dissipate energy. One is to create high velocity and therefore high friction loss or to use the orifice principle to create a local loss. By dividing the flow into small passages and putting multiple orifices in series, the cavitation performance is greatly improved and the noise of the valve much reduced. The most frequently used is multiple losses in series. This can be created with a tortuous flow path or by placing multiple orifices in series. This has been accomplished by specially designed internal trim which can often be inserted into many standard valve bodies. A variation of this principle it to use a single-stage but with multiple orifices. This is like the cavitation trim shown in Fig. 4.1, which is an example of cavitation trim placed inside a standard globe-valve body.

A variation of this same concept is shown where the flow passage is made up of a stack of disks in which complicated flow paths are machined. Each change of direction acts similarly to an orifice and drops the pressure in stages. Properly designed, extremely high pressure drops can be taken with such a valve with virtually no noise or cavitation erosion damage.

### Sleeve Valves

Another type of cavitation control valve is the sleeve valve. It comes in either a straight or angled body. It has a cylindrical body with many small orifices and a second sleeve that slides past the orifices to vary the number exposed to the flow. The reason this type of valve reduces damage caused by cavitation is that the cavitation occurs around the small jets, away from the solid boundary, where it is less likely to cause erosion damage. The valve can operate successfully at high heads. Another style, called the poly-jet valve, can be used in-line and functions like a conventional control valve. The major limitation is that it can only be used with clean liquids.

### Howell–Bunger Valves

The valve shown in Fig. 4.1 is a popular free-discharge valve presently in use. It is commonly used as a turbine bypass valve, for flood control or irrigation, to drain reservoirs, or for aerating water. It is suitable both for low head and high head operation and can function cavitation free under large head drops and high velocities. The valve consists of a fixed cone supported by vanes extending inside the valve body. Flow is controlled by movement of the external sleeve which seats against the fixed cone. When discharging into the atmosphere, the jet spreads out in a wide cone angle and the jet breaks up into a fine spray which helps prevent downstream erosion. The breakup of the jet is efficient in aerating the water.

If containment of the jet is desired, the Howell–Bunger valve can be modified into what is called the ring jet valve; a hood installed around the discharge jet confines the jet to a cylindrical shape, rather than a conical shape. It can also be installed to operate in a submerged condition. For such operation, it is important that the jet be properly aerated to prevent cavitation.

### Hollow-Jet Valves

The hollow-jet valve is a free-discharge valve used fairly extensively for release of water from reservoirs. The forerunner of the Howell–Bunger valve, it shares many of the same characteristics. The main difference is that control is provided with an internal needle that moves upstream against the flow. If properly machined and aerated, it can operate cavitation-free. Since it is more costly to build than the Howell–Bunger valve and has no advantages over it, it has not had much use for a number of years.

### Needle and Tube Valves

These valves were early developments for free discharge controls at reservoirs. There are still some of these valves in service, but they are no longer being made. Information regarding their historical development and use, can be found in Chapter 22 of Reference 14a.

## 4.2   HYDRAULIC CHARACTERISTICS

### Flow Coefficients

The pressure drop across a valve is proportional to the square of the discharge or velocity. The only difference between a valve loss and another minor loss is that the loss coefficient varies with valve opening. For certain valves (mainly small valves), there is actually some variation of the coefficients with Reynolds number. However, only for situations in which the head loss across a valve must be very accurately known would this be important.

The relationship between flow and pressure drop can be expressed as a flow coefficient. There are several coefficients used by different engineering groups, the most common ones being

$$K_l = \frac{2g\Delta H}{V^2} = \frac{2g\Delta H A^2}{Q^2} \qquad (4.1)$$

$$C_v = \frac{Q}{\sqrt{\dfrac{\Delta P}{\text{sg}}}} \qquad (4.2)$$

$$C_{d1} = \frac{V}{(2g\Delta H)^{0.5}} \qquad (4.3)$$

$$C_d = \frac{V}{(2g\Delta H + V^2)^{0.5}} \qquad (4.4)$$

$$C_{df} = \frac{V}{(2gH_u)^{0.5}} \quad \text{(for free-discharge valves)} \qquad (4.5)$$

in which sg = specific gravity of the fluid, $\Delta H$ and $\Delta P$ are, respectively, the total head or pressure drop (including any change in velocity head) caused by the valve at an average velocity $V$ or flow rate $Q$ and $H_u$ is the pressure head upstream from the valve. Note that for a free-discharge valve $H_u = \Delta H$, since $H_d = 0$. All of the equations but Eq. 4.2 are dimensionless, and therefore, any consistent system of units can be used. In the U.S. system, the customary units are $g = 32.2$ fps$^2$, $\Delta H$ in ft, $V$ in fps, and $Q$ in cfs. In SI units, $g = 9.81$ m/s$^2$, $\Delta H$ in meters, $V$ in m/s, and $Q$ in m$^3$/s. Equation 4.2 is not dimensionless, and for units, $Q$ is in gpm and $\Delta P$ in psi. This equation is used extensively in the waterworks field. It represents the flow rate in gpm that a valve can supply at a pressure drop of 1 psi.

When comparing the flow coefficients of identical valves of different sizes, $K_l$, $C_d$, and $C_{d1}$ will be the same (ignoring any Renolds number effects). $C_v$ can be scaled from a valve of diameter $d_2$, to one of diameter $d_1$ using

$$C_{v1} = C_{v2}\left(\frac{d_1}{d_2}\right)^2 \qquad (4.6)$$

To understand and use these coefficients properly, some discussion about $V$ and $\Delta H$ is necessary. Since the coefficients are experimentally determined, one must have some understanding of the method used and what factors influence the coefficients. Factors influencing these coefficients and how they are used include the type of valve (whether it is in-line or free-discharge) and if the pipe is the same diameter upstream and downstream and if it is the same as the valve diameter.

First, the magnitude of any of the coefficients (Eqs. 4.1–4.4) applies only to the valve when it is installed in the same diameter pipe as used for the tests. The pipe diameter used for testing is usually the same as the valve diameter. If a valve is used in a pipe of constant diameter upstream and down, the $\Delta P$ or $\Delta H$ in Eqs. 4.1–4.4 is both the net pressure drop and net loss of total energy. If it is installed in a larger pipe with smooth transitions, the experimental coefficient applies if the velocity is based on the valve diameter rather than the pipe diameter. If the valve is installed in a larger pipe, and especially if it is installed with reducing flanges, $\Delta H$ will be larger and will reduce the flow coefficients. If the upstream and downstream pipes are of different diameters, $\Delta H = \Delta P/\gamma + \Delta(V^2/2g)$. Also, if there are disturbances upstream from the valve, the coefficients are affected.

When a valve is used for free discharge, the discharge coefficient is reduced because there is no pressure recovery downstream. Equation 4.5 is the equation traditionally used for free-discharge valves.

For full-ported valves, like ball or cone valves, where the net head drop for the valve fully open is almost zero, it is best to use Eq. 4.4. For this case, $C_v$ and $C_{d1}$ both go to infinity, whereas $C_d$ only varies between 0 and 1.0.

To use the coefficients properly, it is helpful to understand the method of evaluating $\Delta P$ and $\Delta H$. To explain this problem, it is necessary to discuss energy dissipation and the general flow conditions at a minor loss.

Figure 4.2 shows the energy grade line (EGL) and hydraulic grade line (HGL) for flow through an orifice. The principles to be discussed apply equally to any minor loss. At point 2 (called the vena contracta), the jet diameter is minimum and the velocity maximum. The pressure drop between points 1 and 2 is due to acceleration of the fluid. For an orifice, if the contraction coefficient is known, P1 − P2 can be estimated by the energy equation, assuming no losses. Between points 2 and 3, the jet is dissipated by viscous shear and the pressure increases. The momentum equation can be used to calculate P3. At point 3, the velocity profile is reasonably uniform and the pressure recovery complete. Beyond point 3, a normal hydraulic grade line is established which is parallel to the HGL upstream if the pipe has a constant diameter.

The pressure drop caused by an orifice or valve is caused by the generation and dissipation of turbulent eddies in the high shear layer around the submerged jet. It is measured by installing piezometers upstream and

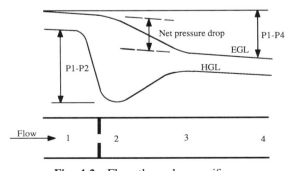

**Fig. 4.2**  Flow through an orifice.

downstream. Their placement is important. The upstream piezometers can be placed one or more pipe diameters upstream. The downstream piezometer must be placed so it is beyond the zone of flow establishment (the region where the pressure is increasing with distance). Referring to Fig. 4.2, this could be any position beyond point 3. If it is placed between points 2 and 3, an artificially high $\Delta P$ will be measured. The distance to the point where the pressure recovery is complete varies with the type of valve or minor loss. The distance is at least 5–8 pipe diameters. Consider that points 1 and 4 are selected. The measured P1 − P4 is shown graphically as the total drop in the EGL. The magnitude of P1 − P4 increases as point 4 is located farther downstream.

Friction losses in the pipe should not be included in determining $\Delta H$ or $\Delta P$ for Eqs. 4.1–4.4. The accepted procedure for determining the net pressure drop is to subtract line loss. This is demonstrated in Fig. 4.2 by projecting the two EGL lines. The net loss is the distance between them, labeled net pressure loss ($\Delta H_{net}$). This is calculated by

$$\Delta H_{net} = \Delta H_{measured} - \frac{fLV^2}{2gd}$$

in which $f$ is measured before the device is installed and $L$ is the distance between points 1 and 4.

The importance of this is best demonstrated by an example. Consider the following experimental data taken on a 24-in. butterfly valve fully open (64). The valve was installed in a straight section of 24-in, ID pipe with piezometers located 1 diameter upstream and 6.5 diameters downstream. For one test point, the measured data were: $\Delta P = 0.355$ psi, $Q = 21,000$ gpm. The friction factor for the pipe had been previously measured as 0.012. The net pressure drop was calculated to be 0.227 psi. The flow coefficient $C_v$ for the two $\Delta P$ values are

$$C_v = 21,000/(0.355)^{0.5} = 35,200$$

$$C_v = 21,000/(0.227)^{0.5} = 44,100$$

The difference is about 20%. This particular valve had an unusually high $C_v$ because of the disk design. The problem of using the measured instead of the net $\Delta P$ is more significant at large valve openings.

Another factor influencing the flow coefficients and the performance of valves is the upstream and downstream piping. If an elbow, tee, or other minor loss is located too closely, the approach velocity profile and turbulence can change the flow coefficient, torque, cavitation, and stability of the valve. A downstream disturbance can affect the pressure recovery.

Since there are several coefficients in use, the following conversions are included to aid in transferring from one definition to another.

$$K_l = \frac{1}{C_d^2} - 1 = \frac{1}{C_{d1}^2} \tag{4.7}$$

$$C_{d1} = \left(\frac{1}{K_l}\right)^{0.5} \tag{4.8}$$

$$C_d = \left(\frac{1}{(K_l + 1)}\right)^{0.5} \tag{4.9}$$

$$C_{d1} = \left[\frac{C_d^2}{(1 - C_d^2)}\right]^{0.5} \tag{4.10}$$

$$C_d = 1 \bigg/ \left(\frac{890d^4}{C_v^2} + 1\right)^{0.5} \quad (d \text{ in inches}) \tag{4.11}$$

$$C_{d1} = \frac{C_v}{(\text{sg}^{0.5} 29.84 d^2)} \quad (d \text{ in inches}) \tag{4.12}$$

$$C_v = \left[\frac{890d^4 C_d^2}{(1 - C_d^2)}\right]^{0.5} \quad (d \text{ in inches}) \tag{4.13}$$

Values of the flow coefficients vary with type of valve and for the same valve due to minor changes in design. It is therefore necessary to obtain information from the valve manufacturers on the specific valve of interest. For general information and use in example problems, representative data for several valves are presented in Figs. 4.3 and 4.4 for in-line and free discharge valves, respectively.

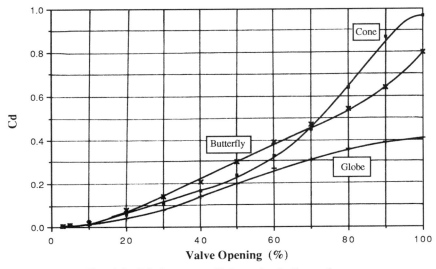

**Fig. 4.3**   Discharge coefficients for in-line valves.

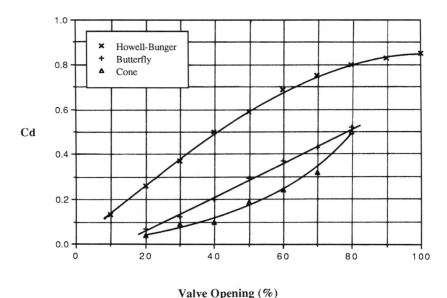

**Valve Opening (%)**

**Fig. 4.4**   Discharge coefficients for free-discharge valves.

### Valve Versus System Loss

When selecting a control valve, it is necessary to analyze its performance as a part of the piping system and not consider it an isolated device. For example, valve manufacturers often transform the flow coefficient data into a curve of percent maximum flow versus valve opening. Such information can be misleading because the controllability of a valve depends on the system in which it is installed. The same valve installed in different systems will have totally different percent flow versus valve opening characteristics. This is demonstrated in Example 4.1.

***Example 4.1.*** Calculate discharge versus valve opening for flow between two reservoirs controlled by a butterfly valve for (a) a short pipe where friction is small, $fL/d = 3$, and (b) a long pipe with high friction, $fL/d = 250$).

*Case A.* Write the energy equation between the reservoirs assuming an elevation difference of 30 ft. Assume minor losses are negligible. Therefore, $30 = (3 + K_l)V^2/2g$. ($K_l$ is for the valve). When the butterfly valve is fully open, $C_d = 0.80$ (Fig. 4.3) and $K_l = 0.563$ (Eq. 4.7). So $V = 23.3$ fps.

As the valve closes the velocity is reduced as shown below:

| VO (%) | $C_d$ | $K_l$ | $V_{fps}$ | % of $V_{max}$ |
|--------|-------|-------|-----------|----------------|
| 90 | 0.64 | 1.44 | 20.9 | 89 |
| 52 | 0.32 | 8.77 | 13.2 | 56 |
| 30 | 0.14 | 50.0 | 6.04 | 26 |
| 15 | 0.06 | 277 | 2.62 | 11 |

*Case B.* With high friction losses ($fL/d = 250$ and ignoring minor losses, except for the valve), the energy equation gives $\Delta H = (250 + K_l)V^2/2g$. For this case, a higher $\Delta H$ is used, so a more reasonable velocity at full open is obtained. Let $\Delta H = 200$ ft. So $V = 7.17$ fps.

As the valve closes, the flow changes as listed below:

| VO (%) | $C_d$ | $K_l$ | V (fps) | % of $V_{max}$ |
|--------|-------|-------|---------|----------------|
| 90 | 0.64 | 1.44 | 7.16 | 99.8 |
| 52 | 0.32 | 8.77 | 7.06 | 98.4 |
| 30 | 0.14 | 50.0 | 6.55 | 91.3 |
| 15 | 0.06 | 277 | 4.94 | 68.9 |
| 10 | 0.02 | 2500 | 2.16 | 30.0 |

Case A shows that the flow decreases almost linearly as the valve is closed. At 52% open, the flow is less than 56% of its original value. For Case B, the situation is different. Because line loss is high and $fL/d \gg K_l$, closing the valve has little influence on the flow for about the first 70% of valve closure. As a result of the system–valve interaction, it is misleading to think of a valve as having a linear or nonlinear closing characteristic.

For the system described as Case B, if it were desired to have a valve with a wider range of control, one could either use a smaller butterfly valve or a globe valve. First, consider a smaller butterfly valve. Assume that it is desired to limit the head loss across the full open valve to 5% of the total head loss; so $K_l = 0.05(250 + K_l)$, $K_l = 13.2$. Find the required size of butterfly valve using subscript 1 as conditions in the pipe and 2 for the valve. Note that $K_{l1} = 13.2$ is based on $V_1^2/2g$ ($V_1$ = average pipe velocity). Also note that $K_{l2} = 0.563$ for any size of geometrically similar butterfly valve fully open (based on Fig. 4.3) so $\Delta H = 0.563Q^2/2gA_2^2 = 13.2Q^2/2gA_1^2$ which gives $d_2 = 0.454d_1$.

If a globe valve were used, $C_d = 0.41$ (Fig. 4.3 for a full open valve), $K_{l2} = 4.95$ and $K_{l1} = 13.2$, so $d_2 = 0.782\,d_1$.

The range of opening over which the valve controls the flow is important for sizing pressure reducing valves and determining the safe closure time for control valves, both of which will be discussed subsequently.

### Torque Coefficients

Forces required to open or close quarter-turn valves (valves that rotate 90° like butterfly and cone valves) are caused by friction and hydrodynamic forces. Friction forces act at the valve seat and bearing surfaces. They can be significantly increased due to the presence of fluid forces. For example, the pressure drop across a valve causes a significant load on the shaft, which increases the bearing friction. Pressure can also distort rubber seats and influence the seating torque. For small butterfly valves, the maximum torque is usually the seating or unseating torque. The magnitude of these forces is significantly altered by the design of the leaf and seat. Valves with metal or Teflon seats and eccentric shafts will have very little seating torque. The other extreme is the symmetrical butterfly valve that has a body and leaf which are fully coated with rubber. This design often leads to large seating torques.

Hydrodynamic torque $(T)$ is caused by forces induced by the flowing water. This torque can be related to the valve diameter $(d)$ and velocity $(V)$ or pressure drop $(\Delta P)$ by the following equations:

$$C_{tv} = \frac{T}{\rho d^3 V^2} \qquad (4.14)$$

$$C_{tdP} = \frac{T}{d^3 \Delta P} \qquad (4.15)$$

These torque coefficients are dimensionless if a consistent set of units is used. For a sign convention, a positive value of the torque coefficient represents a torque that tries to close the valve.

Torque coefficients can be misleading if not completely understood. To help explain some of these potential misunderstandings, consider the $C_{tdP}$ data for a cone valve shown in Fig. 4.5. The maximum torque coefficient occurs at about 83°, which, however, does not mean that the torque is maximum at 83°. Since $\Delta P$ varies significantly with valve opening, rapidly increasing as the valve is closed, the maximum hydrodynamic torque occurs at a smaller valve opening even though the $C_{tdP}$ is smaller. When sizing an operator, one must consider the full range of valve operation to determine the maximum torque (including friction and seating torque).

To demonstrate another problem, compare the two curves in Fig. 4.5. For the 12-in. valve installed in a 15-in. pipe using reducing flanges, the maximum $C_{tdP}$ is reduced by about 50%. One might be tempted to conclude that the hydrodynamic torque is likewise reduced. Such is not the case, as shown in Fig. 4.6. Using $C_{tv}$ data not shown demonstrates that the valve installed in either the 12-in. or 15-in. pipes has the same torque coefficient

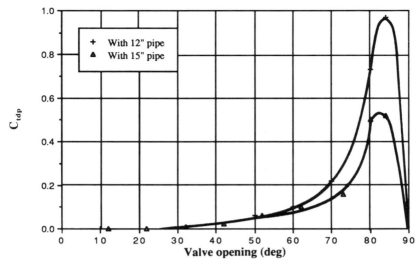

**Fig. 4.5** Torque coefficients for a cone valve.

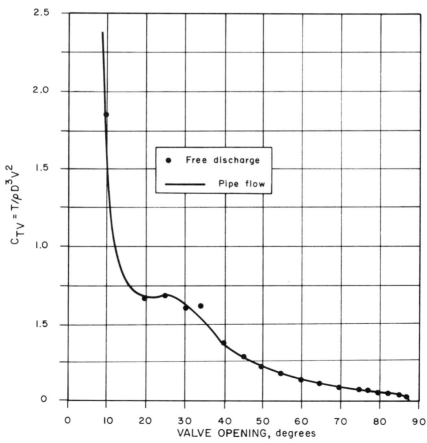

**Fig. 4.6** Torque coefficients for a cone valve.

for the same flow rate ($V$ = velocity based on a 12-in. valve diameter for both cases). What happened is that the $\Delta P$ increased due to the sudden changes in pipe diameter at the reducing flanges. This reduced $C_{tdP}$ even though the torque was not changed.

A similar problem is observed when the valve is operated in a free discharge condition. Free-discharge means that either there is no downstream piping or that the downstream pipe is vented to the atmosphere. Figure 4.6 shows that there is no change in $C_{tv}$ for free-discharge versus normal pipe flow. If the comparison were made based on $C_{tdP}$, there would be a decrease in the coefficient because $\Delta P$ is larger for free discharge since there is no pressure recovery.

For butterfly valves, large differences in the torque are possible by minor modifications to the shape of the disk, by having the shaft slightly eccentric, or by installing the valve backward. Figure 4.7 shows torque data for a 24-in. butterfly (64). The two sets of data (referred to as valves A and B) are for the same valve operating in different quadrants of rotation. Valve B appears to have a superior torque characteristic. The magnitudes of $C_{tdP}$ are less and there is no torque reversal. However, the $C_v$ for valve B is

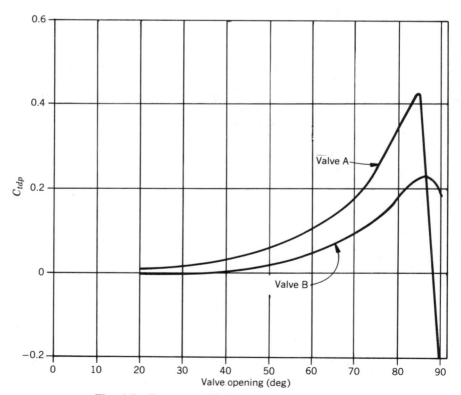

**Fig. 4.7**  Torque coefficients for a butterfly valve.

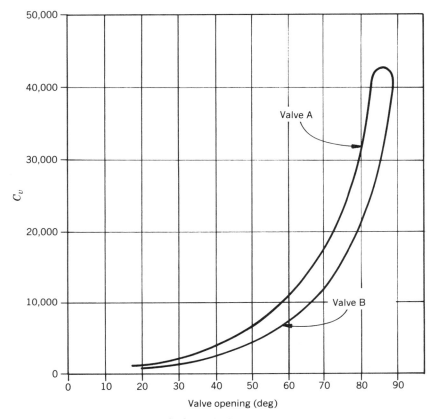

**Fig. 4.8** $C_v$ for a 24-in. butterfly valve.

considerably less than valve A at each valve opening (see Fig. 4.8). Therefore, part of the reduction in $C_{tdP}$ is due to a larger $\Delta P$ and not necessarily a smaller torque. To make a complete evaluation, one must consider both the torque coefficient and the flow coefficient data. This is demonstrated in the next example.

***Example 4.2.*** A 24-in. butterfly valve whose torque and flow coefficients are shown in Figs. 4.7 and 4.8, respectively is used to control flow between two reservoirs. The friction and minor losses are equivalent to $fL/d + \Sigma K_l = 10$ (excluding the valve). Calculate the torque on the valve at 40, 60, and 84° rotation for valves A and B. Assume that the difference in reservoir elevations is 120 ft.

First calculate the flow rate and pressure drop across the valve at each valve setting. The energy equation applied between the two reservoirs gives $120 = (10 + K_l)\ V^2/2g$

Obtain $C_v$ values from Fig. 4.8 and calculate $K_l$ for the valve with Eqs. 4.12 and 4.7 (or Eqs. 4.11 and 4.7). For valve A at $84°$, $C_v = 40,500$, $C_{tdP} = 0.43$, $C_{d1} = C_v/(29.84\ D^2) = 2.36$, and $K_l = 1/C_{d1}^2 = 0.180$.

$V = [2g(120)/10.18]^{0.5} = 27.5$ fps, $Q = 27.5$ fps $\cdot\ 3.14$ ft.$^2 \cdot 448.8$ gpm/ cfs $= 38,860$ gpm. For the valve $\Delta P = (Q/C_v)^2 = (38,860/40,500)^2 = 0.921$ psi $= 133$ psf.

The flow torque is (from Eq. 4.15) $T = C_{tdP}d^3\ \Delta P = 0.43\ (2\ \text{ft})^3\ (133$ psf$) = 456$ ft-lb. Torques for the other conditions are summarized below:

| Valve Opening (deg) | $C_v$ | $C_{tdP}$ | $K_l$ | $V$ (fps) | $Q$ (gpm) | $\Delta P$ (lb/ft$^2$) | $T$ (ft-lb) |
|---|---|---|---|---|---|---|---|
| | | | *Valve A* | | | | |
| 84 | 40,500 | 0.43 | 0.18 | 27.5 | 38,800 | 133 | 456 |
| 60 | 10,500 | 0.11 | 2.68 | 24.7 | 34,800 | 1,580 | 1,390 |
| 40 | 3,400 | 0.035 | 25.5 | 14.8 | 20,800 | 5,390 | 1,510 |
| | | | *Valve B* | | | | |
| 84 | 28,500 | 0.23 | 0.364 | 27.3 | 38,500 | 263 | 484 |
| 60 | 7,000 | 0.05 | 6.03 | 22.0 | 31,000 | 2,800 | 1,120 |
| 40 | 2,500 | 0.01 | 47.3 | 11.6 | 16,300 | 6,121 | 490 |

The results of Example 4.2 show that even though the $C_{tdP}$ for valve A at $84°$ is twice as large, the actual torque is less than that for valve B. By making these calculations at all valve openings, the maximum hydrodynamic torque can be evaluated. Note the large difference in torque at $40°$. For valve A, the torque is about three times that for valve B. This is typical for butterfly valves because of large differences in disk design. This emphasizes the need to make torque calculations at all valve openings. Use of modern spreadsheet applications makes the calculations easy. When considering the total torque required to operate the valve, bearing friction and seating torque must also be included.

Another important aspect of torque is flutter. This is caused by rapid variations in torque due to instabilities in the flow. The butterfly valve is especially susceptible to flutter because it acts much like a flag in the wind at large openings. This problem is worse when there is torque reversal. For example, refer to valve A in Fig. 4.7. Up to $84°$, torque tries to close the valve. Between 84 and $90°$, the torque coefficient changes from 0.42 to $-0.19$. Above about $87°$, the torque tries to open the valve. When the valve operates in this range, the leaf will flutter at high flow velocities. With time, the flutter causes wear which further increases the flutter and accelerates the wear. It is best not to operate valves near points of torque reversal.

Even if there is no torque reversal, the torque is always unstable and causes some flutter for valves that operate with a rotational motion. How severe the flutter is depends on the flow conditions approaching the valve, the type of operator, the amount of play in the couplings, the strength and fatigue characteristics of the shaft, connections and couplings, and whether or not there is any cavitation.

The approach flow conditions have a great effect on flutter. The author inspected a water treatment plant in which the gear operator on a 24-in. butterfly valve was continually vibrating loose and almost fell off the valve. It was installed as an isolation valve one pipe diameter downstream from a similar valve used for control. The high velocity jets from the control valve caused the leaf of the isolation valve to flutter violently. Not only is the flutter increased but the magnitude of the mean torque can be larger if the approach flow is disturbed. It is best not to place a control valve just downstream from any major disturbance.

The type of operator and play in the couplings can increase the movement of the disk or plug during flutter. For example, if a pneumatic operator is used, the compressibility of the air will allow more movement.

## 4.3   VALVE SELECTION AND SIZING

When selecting and sizing control valves, the following principles should be considered:

1. Select the type and size that do not produce excessive loss when fully open.
2. Be sure that the valve controls the flow over at least 50% of its stroke.
3. Avoid excessive cavitation.
4. Specify an operating procedure, including a closing time, that will not produce undesirable transients.
5. Do not operate at a valve opening below about 10–15% open.
6. Some valves do not control near full open and should not be operated above about 90% open.

### Valve loss

Unless a valve is primarily used for energy dissipation, it is usually desirable to have a low pressure drop when it is full open. For many pumping and power-generating systems, the loss across a valve represents a great deal of power wasted over the life of the project. Therefore, there can be a significant cost savings by reducing the valve loss. Loss coefficients vary significantly with valve type and design. Globe valves typically have $K_l$

values, between 4 and 6 at full open. The loss coefficient for full-ported cone and ball valves is almost zero. Butterfly valves can vary from about 0.6 to 0.1, depending on the size and leaf design.

With low-loss valves such as cone, ball, and gate, and to some extent the butterfly, it is sometimes desirable to use a valve that is smaller in diameter than the pipe to have better control. Conical transitions should be used for reducing and expanding the flow to minimize losses. If the valve is for energy dissipation, reducing the valve diameter increases the cavitation problem. One must therefore balance the advantages of increased controllability with increased cavitation when using a valve that is smaller than the pipe.

### Limiting Valve Opening

The maximum range of openings over which many valves can accurately and safely control flow is generally between about 10 and 90% open. At smaller valve openings, it may be difficult to control flow because a small change in opening causes a large change in $K_l$. Also, at small openings the head drop often becomes large, and for some valves, seat damage can occur by the high velocity jets. At large openings, the valve loss is so small compared with the friction loss of the pipe that the flow does not change significantly with valve opening. This does not apply to all valves. The sleeve valve and globe valves with cavitation trim can operate at very small openings without difficulty. These types of valves increase flow by exposing more small passages rather than just increasing the size of one passage. At large opening these types of valves eventually lose control if too many holes are exposed.

### Flow Control

As a general guide, it is recommended that a control valve be able to control the flow over at least 50% of its movement. This means that as a value is closed to 50% open, the valve should be able to reduce the flow by at least 10%, or that about 15% of the total available head loss is caused by the valve. Example 4.1 clearly demonstrated this. For Case B, a smaller valve would be required if it were to function as a control valve. For that case, using a valve considerably smaller than the pipe would not significantly reduce the maximum flow but would improve controllability.

If a wide range of discharge is necessary, it is recommended that parallel valves of different diameters be considered. Parallel valves are also desirable for reliability. Consider one example, where a 16-in. pipe operating at high pressure was installed to service a developing area. The initial minimum flow was 1 gpm, and the maximum future demand was 10,000 gpm. A valve structure was built with a 3/4-in., 3-in., 6-in., and two 14-in. valves in parallel (89).

## Cavitation

One of the major causes of valve replacement is cavitation damage. Valves can be damaged in a few weeks if operated in heavy cavitation. As the ratio of the pressure drop across a valve to the mean pressure increases, the chance of cavitation increases. Cavitation produces noise, vibrations, and erosion damage and can limit the flow capacity of a valve. It is one of the serious problems related to valve operation.

Each valve has different cavitation characteristics, just as each has different flow and torque characteristics. Cavitation is a somewhat complex subject and is discussed in detail in Chapters 5–7. Cavitation information for several common types of valves, orifices, and other pipe components are presented in Chapters 6 and 7. Procedures and examples for design and analysis are also included.

## Transient Control

Operation of any valve can cause transient; even if it is just a gradual change from one steady-state condition to another. The magnitude of the pressure generated during a transient is directly related to the rate of change of the flow by the valve, the pipe length, and the pipe material. Such pressures can easily exceed the safe operating pressure of a pipe or valve and cause rupture. Two common causes of transients are improper valve operation and pump shutdown. A thorough discussion of transients associated with valve operation is presented in Chapters 8–10, including computational techniques and examples predicting the magnitude of the transients. This section discusses how transients are generated by control valves and how they can be controlled by proper design and operation. Simple examples are presented. Transient information for pressure-relief valves, pressure-reducing valves, and air valves is considered in subsequent sections of this chapter.

Keeping transients caused by operating a control valve to an acceptable limit requires specifying operating procedures, including the safe closing time. The primary factors influencing the safe closing time are the initial velocity, pipe diameter, wall thickness and length, pipe material, and valve characteristics.

The magnitude of the transient pressure $\Delta H$ caused by rapidly reducing the velocity by $\Delta V$ is calculated by the equation

$$\Delta H = -\mathbf{a}\Delta V/g \qquad (4.16)$$

in which $\Delta V = V_2 - V_1$, $V_1$ is the initial velocity, $V_2$ is the velocity after the change of valve opening, $\mathbf{a}$ is the acoustic wave speed, and $g$ is acceleration of gravity. For a steel pipe $\mathbf{a} \sim 3200$ fps, so $\Delta H \sim 100 \ \Delta V$. For a sudden decrease in velocity of only 1 fps, the transient head rise is about 100 ft.

The wave speed $\mathbf{a}$ is a function of the bulk modulus of the liquid, pipe diameter, wall thickness, and the pipe material (assuming there is no trapped air in the pipe). An equation for calculating **a** is derived in Chapter 8.

The normal way to control the transient is to control $\Delta V$. This is done by slowly closing the valve. How slow a valve must be closed depends on the pipe length and wave speed. One effect of pipe length on the safe closing time of a valve can be demonstrated by referring to Example 4.1. For Case A (short pipe), the flow started to reduce as soon as the valve began to close. For the long pipe (Case B), the flow did not significantly change until the valve was over half closed. This means that during the first half of the valve closure time, essentially nothing happens to reduce the flow or cause a transient. All of the change of flow occurred during the last half of the valve movement. If the valve for Case B were initially wide open, the actual closure time would have to be more than double the expected closure time because most of the velocity change occurs at small valve openings. Accurate determination of safe valve closing times can only be done by transient analysis techniques discussed in subsequent chapters.

> *Example 4.3.* Select a valve (or valves) for a control structure to furnish flows from 500 gpm at $\Delta P = 50$ psi to 5000 gpm at $\Delta P = 10$ psi. Note: this problem must include a thorough cavitation analysis to prevent erosion damage and excessive noise and vibrations at the valves.
>
> In selecting a valve (or valves) for this service, the six criteria presented at the beginning of Section 4.3 should be considered. This example considers only items 1 and 5. The problem of controllability (criteria #2) was discussed in Example 4.1. Since cavitation is not discussed until Chapters 5–7, criteria #3 will be omitted in this example (see Chapter 6 for cavitation analysis for valves). Criteria #4 for transients is treated in Chapters 8–10.
>
> Criteria #1: At maximum flow (5000 gpm), the allowable head drop is 10 psi across the valve. Since this allowable $\Delta P$ is reasonably large and the required $\Delta P$ increases at smaller flows, see if a globe valve can be used. From Fig. 4.3, $C_d = 0.415$ for the globe valve at full open.

$$\Delta H = 10 \cdot 2.308 = 23.1 \text{ ft}$$

$$K_l = \frac{1}{0.415^2} - 1 = 4.81 \text{ (Eq. 4.7)}$$

$$V = \left(2g\, \frac{\Delta H}{K_l}\right)^{0.5} \text{ (Eq. 4.1)}$$

$$Q = \frac{5000 \text{ gpm}}{448.8} = 11.1 \text{ cfs}$$

$$A = \frac{Q}{V} = 0.633 \text{ ft}^2$$

$$d = 0.898 \text{ ft} = 10.8 \text{ in}$$

Since only standard valve sizes can be used, the next largest size is 12 in., so $A = 0.7854$ ft$^2$, $V = 14.2$ fps, $\Delta H = 15.1$ ft or 6.52 psi. Since the actual $\Delta P$ is less than the allowable of 10 psi, the first criterion is met with a 12-in. globe valve. If a 10-in. globe valve were chosen, the head loss would be slightly greater than 10 psi. For cavitation control, the 12-in. valve would be preferred. For flow control, the 10-in. valve would be a better choice.

Criterion #4 suggests a minimum valve opening of 10 to 15%. At $Q_{min} = 500$ gpm and $\Delta P = 50$ psi (115.4 ft), find the required opening of the 12-in. globe valve. $V = Q/A = 1.42$ fps. Using Eq. 4.4, $C_d = 0.016$. From Fig. 4.3, $C_d$ at 10% is about 0.02, so the 12-in. globe valve will be operating at about 10% at minimum flow, which is in its control range.

***Example 4.4.*** Three 16-in. centrifugal pumps are used to supply water to a 24-in. 15900-ft long pipeline ($f = 0.014$ and $\Delta z = 50$ ft). The pump characteristics for one, two, and three pumps operating are shown in Fig. 4.9. Find the flow and torque on the 16-in. butterfly valves at each 10% open. One of the purposes of this example is to demonstrate how important it is to make calculations for all flow conditions when evaluating the maximum torque to be used in selecting the valve operator. The example shows that the maximum torque occurs when opening a valve with only one pump operating and that this is about triple the torque for three pump operation.

First, evaluate the system equation (Eq. 3.1): $H_p = \Delta z + CQ^2$. Note that $Q$ is the flow in the main pipe and not the flow from one pump

$$C = \left[ \frac{fL}{(2gdA_p^2)} + \frac{K_l}{(2gA_v^2N^2)} \right]$$

$\Delta z = 50$ ft

$A_p = 3.14$ ft$^2$

$K_l =$ valve loss coefficient

$A_v =$ area of valve

$N =$ number of valves or pumps in use

so:

$$H_p = 50 + CQ^2 \tag{4.17}$$

$$C = 8.66^{-7} + 3.95^{-8} \frac{K_l}{N^2}$$

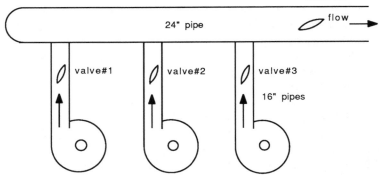

System parameters: L = 15900 ft. D = 24-in. f = 0.014
elevation head = 50 ft.

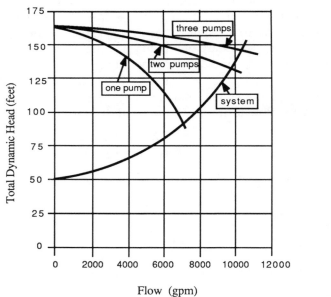

**Fig. 4.9** Pipe system for Example 4.4.

Next, evaluate the equations for the pumps (Eq. 3.3): $H_p = H_0 - C1Q - C2Q^2$ by scaling $H_0$ and values of $H_p$ and $Q$ at two points on the single pump curve. The folowing points are used:

$$H_0 = 165 \text{ ft}$$

$$141 = 165 - 4000C1 - 4000^2C2$$

$$100 = 165 - 6900C1 - 6900^2C2$$

Solving these two equations gives

$$C1 = 0.00116 \text{ and } C2 = 1.2 \times 10^{-6}$$

The *one-pump* equation becomes

$$H_p = 165 - 0.00116Q - 1.2 \times 10^{-6} \tag{4.18}$$

For two pumps in parallel, the equation can be derived by substituting $Q/2$ into Eq. 4.18. This gives $C1 = 0.000586$ and $C2 = 3 \times 10^{-7}$, so the *two-pump* equation becomes

$$H_p = 165 - 0.000586Q - 3 \times 10^{-7}Q^2 \tag{4.19}$$

For three pumps in parallel, the equation can be derived substituting $Q/3$ into Eq. 4.18. This gives $C1 = 0.000387$ and $C2 = 1.33 \times 10^{-7}$ so the *three-pump* equation becomes

$$H_p = 165 - 0.000387Q - 1.33 \times 10^{-7}Q^2 \tag{4.20}$$

The flow is determined by solving Eq. 4.17 with each of the pump equations (4.18–4.20). The general form of the solution is

$$Q = 0.5[-B + SQRT(B^2 + 4D)] \tag{4.21}$$

where $B = C1/(C + C2)$ and $D = 115/(C + C2)$ and the $C$, $C1$, and $C2$ values are those in Eqs. 4.17–4.20.

The problem is solved with a spreadsheet program because of the large number of calculations. Table 4.2 lists the output. The other input data required are values for the discharge and torque coefficients. Typical values are listed in columns 2 and 10 of Table 4.2.

**TABLE 4.2  Data for Example 4.4**

| Valve Opening, (%) | Valve, $C_d$ | Valve, $K_l$ | Constant, $C$ in Eq. 3.1 | Constant, $C_1$ in Eq. 3.3 | Constant, $C_2$ in Eq. 3.3 | Total Flow, gpm | Pump Flow, gpm | Pressure Drop, ft | Torque Coef $C_{c\iota\Delta p}$ | Flow Torque ft-lb |
|---|---|---|---|---|---|---|---|---|---|---|
| 1 | 2 | 3 | 4 | 5 | 6 | 7 | 8 | 9 | 10 | 11 |
| One pump | | | | | | | | | | |
| 10 | 0.010 | 9999.0 | 0.0004 | 0.00116 | 1.2E-06 | 537 | 537 | 114 | 0.002 | 33 |
| 20 | 0.090 | 122.5 | 5.7E-06 | 0.00116 | 1.2E-06 | 3998 | 3998 | 77 | 0.010 | 114 |
| 30 | 0.160 | 38.1 | 2.4E-06 | 0.00116 | 1.2E-06 | 5516 | 5516 | 46 | 0.028 | 188 |
| 40 | 0.220 | 19.7 | 1.6E-06 | 0.00116 | 1.2E-06 | 6160 | 6160 | 29 | 0.042 | 182 |
| 50 | 0.340 | 7.65 | 1.2E-06 | 0.00116 | 1.2E-06 | 6728 | 6728 | 14 | 0.085 | 171 |
| 60 | 0.430 | 4.41 | 1E-06 | 0.00116 | 1.2E-06 | 6911 | 6911 | 8 | 0.140 | 171 |
| 70 | 0.520 | 2.70 | 9/7E-07 | 0.00116 | 1.2E-06 | 7013 | 7013 | 5 | 0.220 | 169 |
| 80 | 0.600 | 1.78 | 9.4E-07 | 0.00116 | 1.2E-06 | 7071 | 7071 | 4 | 0.300 | 155 |
| Two pumps | | | | | | | | | | |
| 10 | 0.010 | 9999.0 | 1E-04 | 0.00058 | 3E-07 | 1070 | 535 | 113 | 0.002 | 33 |
| 20 | 0.090 | 122.5 | 2.1E-06 | 0.00058 | 3E-07 | 6837 | 3419 | 57 | 0.010 | 83 |
| 30 | 0.160 | 38.1 | 1.2E-06 | 0.00058 | 3E-07 | 8450 | 4225 | 27 | 0.028 | 110 |
| 40 | 0.220 | 19.7 | 1.1E-06 | 0.00058 | 3E-07 | 8984 | 4492 | 16 | 0.042 | 97 |
| 50 | 0.340 | 7.65 | 9.4E-07 | 0.00058 | 3E-07 | 9394 | 4697 | 7 | 0.085 | 83 |
| 60 | 0.430 | 4.41 | 9.1E-07 | 0.00058 | 3E-07 | 9514 | 4757 | 4 | 0.140 | 81 |
| 70 | 0.520 | 2.70 | 8.9E-07 | 0.00058 | 3E-07 | 9579 | 4790 | 2 | 0.220 | 79 |
| 80 | 0.600 | 1.78 | 8.8E-07 | 0.00058 | 3E-07 | 9615 | 4808 | 2 | 0.300 | 71 |
| Three Pumps | | | | | | | | | | |
| 10 | 0.010 | 9999.0 | 4.5E-05 | 0.00039 | 1.33E-07 | 1596 | 532 | 112 | 0.002 | 33 |
| 20 | 0.090 | 122.5 | 1.4E-06 | 0.00039 | 1.33E-07 | 8526 | 2842 | 39 | 0.010 | 57 |
| 30 | 0.160 | 38.1 | 1E-06 | 0.00039 | 1.33E-07 | 9766 | 3255 | 16 | 0.028 | 65 |
| 40 | 0.220 | 19.7 | 9.5E-07 | 0.00039 | 1.33E-07 | 10117 | 3372 | 9 | 0.042 | 54 |
| 50 | 0.340 | 7.65 | 9E-07 | 0.00039 | 1.33E-07 | 10368 | 3456 | 4 | 0.085 | 45 |
| 60 | 0.430 | 4.41 | 8.9E-07 | 0.00039 | 1.33E-07 | 10438 | 3479 | 2 | 0.140 | 43 |
| 70 | 0.520 | 2.70 | 8.8E-07 | 0.00039 | 1.33E-07 | 10476 | 3492 | 1 | 0.220 | 42 |
| 80 | 0.600 | 1.78 | 8.7E-07 | 0.00039 | 1.33E-07 | 10497 | 3499 | 1 | 0.300 | 38 |

Each column of the spreadsheet contains the following data or equation:

Columns 1 and 2 are given values for valve opening and $C_d$
Column 3 transforms $C_d$ to $K_l$ with Eq. 4.7
Columns 4, 5 and 6 are constants from Eqs. 4.17 − 4.20 (or Eqs. 3.1 and 3.3)
Column 7 is the solution of Eq. 4.21
Column 8 = col. 7 divided by the number of pumps operating
Column 9 is the pressure drop across the valve
Column 10 is the assumed values for the torque coefficients
Column 11 is the torque

With one pump operating and the valve opened, high flow rates and torques occur at small openings. The maximum torques occur at 30% open. The results are summarized below:

| valve No. of Pumps | Flow (gpm) | Torque (ft-lb) |
|:---:|:---:|:---:|
| 1 | 5516 | 188 |
| 2 | 4225 | 110 |
| 3 | 3255 | 65 |

The maximum torque for one pump is almost three times the torque for three pumps. If the analysis were only done for three pump operation the valve operator would be undersized.

## 4.4  PRESSURE REGULATING VALVES

### Pressure-Relief Valves

These valves are often globe-style, activated automatically by line pressure. Their purpose is to open rapidly if the pressure exceeds a predetermined value and close slowly. Such valves can create more problems than they solve if they are not designed and operated properly. The worst case is an oversized valve that opens and closes rapidly. When an overpressure is sensed, the valve opens. If the valve is too large, the pressure drops below the desired value and the valve closes. If the closure is rapid, it generates a transient pressure rise which causes the valve to again open and so on. The author has tested valves such as this that chattered (opened and closed at a high frequency) and endangered the test pipeline. This problem can be eliminated by modifying the valve so it opens fast but closes slowly and being sure it is not oversized for the system.

The closing time depends on the pipe length and wave speed. Evaluating the closing time follows the same procedure as for control valves. Sizing pressure-relief valves based on steady state operation can be done easily with hand calculations. To predict the transient as the relief valve opens or closes can only be done accurately with a transient computer program that analyzes the system. Such programs are discussed in later chapters. The following examples demonstrate a method of estimating the required valve size based on steady state operation.

***Example 4.5.*** A pressure relief valve is required to protect a penstock against transient pressures caused by rapid closure of the turbine wicket gates during load rejection. The penstock is 2 m in diameter and the normal velocity is 5 m/s. The operating head is 80 m. Select a relief valve that will limit the head to 160 m.

The head rise is related to the velocity change by Eq. 4.16, $\Delta H = -\mathbf{a}\,\Delta V/g$. The allowable velocity change that will produce a 80-m head rise is $\Delta V = -80 \cdot 9.81/1200 = -0.65$ m/s.

The velocity in the penstock after the wicket gates are closed must not be less than $5.00 - 0.65 = 4.35$ m/s. The relief valve must therefore discharge 13.7 m³/s, with an upstream head of 160 m and zero pressure downstream. This would cause a severe cavitation problem unless a free-discharge or other special valve were used. Assume that a Howell–Bunger valve is used which has a $C_{df} = 0.85$ fully open (Fig. 4.5). Use $f = 0.014$ and $L = 20$ m.

Minor losses will be included for this case. Assume the following:

$$K_l = 0.3 \text{ at the connection to penstock}$$

$$K_l = 0.24 \text{ for a short radius } 90° \text{ elbow}$$

$$\frac{fL}{d} = 0.014 \cdot \frac{20}{d_2} = \frac{0.28}{d_2}$$

$$K_l = \frac{1}{C_{df}^2} = 1.38 \text{ for the valve (Eq. 4.7)}$$

The pressure at the inlet to the valve $= H_u = K_l Q^2/2gA_3^2$ ($K_l$ and $A_3$ are for the valve). The energy equation applied between the main pipe and the inlet to the relief valve gives

$$\frac{Q^2}{2gA_1^2} + 160 = \left(1 + 0.3 + 0.24 + \frac{0.28}{d_2}\right)\frac{Q^2}{2gA_2^2} + 1.38\,\frac{Q^2}{2gA_3^2}$$

Try a pipe diameter $d_2 = 1$ m and solve for the valve diameter $D_3$. $A_1 = 3.14$ m$^2$, $A_2 = 0.785$ m$^2$ and $Q$ is 13.7 m$^3$/s. Solving the above equation gives $A_3 = 0.414$ m$^2$ or $D_3 = 0.626$ m.

This example ignores the transient generated by operating the valves. The transient generated during an actual gate or valve operation can only be done accurately with a computer program.

Another common application of a pressure relief valve is to protect a pipeline from overpressurization by a pump. Example 4.6 demonstrates how to select a tentative size for this application. Such a procedure may be adequate, but checking the operation with a transient computer program is preferable.

*Example* **4.6.** A pump supplies water to an 8-in. pipeline. Assume that it is desired to select a pressure-relief valve that will keep the line pressure below 75 psi (173 ft) at the discharge side of the pump. The flow through the pressure-relief valve required to keep the pump head at 173 ft is found from the pump curve. Assume that it is 950 gpm. Therefore, select a valve that will discharge at least 950 gpm at 173 ft of head. Select a globe-style valve and use Fig. 4.3 to get its discharge coefficient. At full open, $C_d = 0.41$, $K_l = 1/C_d^2 - 1 = 4.95$ (Eq. 4.7).

Minor losses include:

$$K_l = 0.6 \text{ for connection to main pipe}$$

$$K_l = 2 \cdot 0.24 = 0.48 \text{ for 2 elbows}$$

$$K_l = 1.0 \text{ for pipe exit}$$

Applying the energy equation from the main pipe to the end of the pressure-relief pipe and ignoring elevation changes gives

$$173 + \frac{Q^2}{2gA_1^2} = 0 + (0.6 + 0.48 + 1 + 4.95)\frac{Q^2}{2gA_2^2}$$

$A_1 = 0.349$ ft$^2$ and $Q = 2.12$ cfs. Solving the equation gives $A_2 = 0.0531$ ft$^2$ and $d_2 = 3.12$ in. A 3 in. valve would be a little small to limit the pressure to 75 psi, so a 4-in. valve would be selected.

One aspect of this example that must be mentioned is the problem of choking cavitation. When cavitation becomes so severe that the mean pressure downstream drops to vapor pressure, the valve is choked. The valve acts as if it were free-discharging (into a vacuum) and the discharge coefficient reduces. This means that the valve will not discharge as much flow as calculated. This is solved by selecting a larger valve. Details on how to predict choking cavitation and the reduction in flow are included in Chapter 6.

Since pressure-relief valves usually operate only infrequently, they need periodic servicing. It is sometimes advisable to consider several small valves instead of one larger valve to improve reliability.

### Pressure-Reducing Valves

This is merely a special type of control valve; the same design principles discussed for flow control valves apply. Since these are normally globe valves, one is sometimes faced with the problem of keeping the full open head loss to an acceptable level and yet having an adequate range of controllability. One must either make a compromise on the valve size or consider two parallel lines of different sizes.

Consider the problems resulting from an oversized valve used to maintain a constant downstream pressure. Since it is oversized, it will only control over a limited range of travel. Thus, small movements of the valve plug will cause fairly large changes in $K_l$ and consequently in the downstream pressure. When the downstream pressure drops, the valve opens to compensate. If it moves too far, the pressure will rise above the set point and the valve will have to close. If it closes too far, the pressure drops too low and the valve opens again. This is called "hunting." This unstable valve operation is obviously undesirable. It can usually be solved by decreasing the size of the valve (if the controllable range is too small) or by slowing down the movement of the valve.

The latter solution can cause an additional problem. If there are rapid changes of pressure in a system and the pressure-reducing valves are set to operate slowly, the pressure may exceed safe limits before the valve can respond. It is therefore advisable to consider placing a relief valve near the pressure-reducing valve to protect the line.

Another interesting problem occurs when pressure reducing valves are placed in series. This problem was observed in a gravity flow pipeline. The pipe dropped over 1000 ft in elevation to the point of delivery. To limit the pipe class, three pressure reducing valves were installed along the pipe. The valves had slow operators to prevent "hunting."

Surging occurred in the system which could not be eliminated by adjusting the valve closure speed. Analysis of the system showed that it was unstable. The friction loss was small compared with valve losses, so there was little damping.

To understand in part what happened, consider that the far downstream valve closes a little to regulate the downstream pressure. This sends a positive pressure wave up the pipe. When it reaches the next valve, that valve closes some to compensate for the higher pressure. The closure of the second valve sends a positive pressure wave to the third valve and a negative wave to the first valve. Each time a pressure wave reaches a valve, the valve reacts, trying to keep a constant pressure in the pipe, and the surging continues.

One way that this can be avoided is to have enough friction and other fixed losses to dampen out the wave partially before it reaches the next valve. A better, but more expensive, solution would be to isolate each section of the pipe with a small open water tank.

### Surge-Anticipating Valves

Controllers can be placed on valves so that they automatically open or close when a transient is anticipated. For example, when a pump is shut down, a low pressure wave is sent down the pipe. Under certain conditions, column separation can occur, which may lead to excessive pressures. It is possible to have a valve at the discharge end automatically start closing the instant the pump is turned off. The positive pressure wave from the valve combines with the negative wave from the pump to keep the pressure positive in the pipe.

A second example is a power operated check valve that closes at a controlled rate when the pump is shut off. A third example is an automatic drain valve used in conjunction with a normal check or power-operated check valve. When the pump shuts down and the check valve closes, the drain valve automatically opens to prevent high pressures.

### 4.5   CHECK VALVES

These valves are installed in suction lines of pumps to keep pumps primed or in discharge lines to minimize flow reversal, which can damage pumps and motors. Check valves should open easily for forward flow, firmly seat so they are stable under foward flow conditions, have a low loss coefficient for forward flow, and close when the flow reverses so as to not generate undesirable transients in the pipeline.

With no flow in the system, the valve remains closed or nearly closed due to either gravitational force, a spring-loaded disk, or a combination of the two. As flow is initiated, the dynamic force of the moving water causes the valve to open.

For most styles of check valves, orientation of the installed valve is important, since gravity usually assists the disk in closing. The approach flow conditions are also an important factor affecting the stability of the disk. If an elbow or other disturbance is located a short distance upstream, the unstable flow can cause the disk or float to oscillate, causing surges in the system.

There are numerous designs available, each having a different closing characteristic. The simple swing check valve (Fig. 4.10) is designed with a pivot point above the periphery of the disk. It closes by gravity when the weight of the disk is greater than the dynamic force of the water holding it open. The opening and closing characteristics can be altered by changing axis of rotation, the weight of the disk, or by adding counterweights.

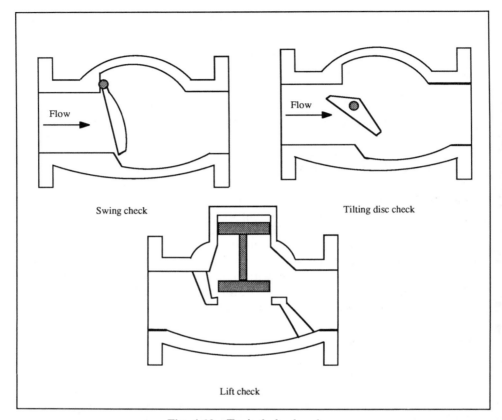

**Fig. 4.10** Typical check valves.

The tilting disk check valve (Fig. 4.10) also depends on gravity for closing, but it has a much shorter distance to travel. Its axis of rotation is just off the center of the disk instead of above the periphery of the disk.

Lift check valves (Fig. 4.10) have floats that raise when the dynamic force exceeds its weight. The rubber flapper and double-door valves are alternative designs of the simple swing check (Fig. 4.11). The flapper does not pivot from a hinge pin, but instead flexes. Also, the seat is on a larger angle, permitting a shorter stroke.

The disk of the double door check-valve design difference is split into equal halves, resulting in shorter strokes for each door. Additionally, each door is forced closed by an internal torsion spring. Other styles are spring-loaded for more rapid closure. Two such valves are the spring-loaded check valve and the nozzle check (Fig. 4.11). The nozzle check valves have exceptionally fast closing characteristics because they have only one moving part with little mass and a short stroke. The streamlined shape of the nozzle offers little resistance to forward flow.

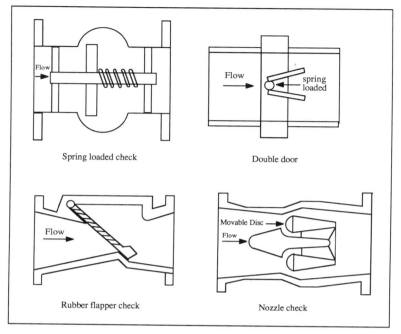

**Fig. 4.11**  Alternative check-valve designs.

Nonreturn valves can also be power operated. Any standard control valve can serve as a nonreturn valve if it is automatically closed whenever reverse flow occurs.

In selecting a nonreturn valve, many compromises must be made, but the two primary requirements are that it has an acceptable loss coefficient for forward flow and that it does not create excessive transient pressures upon closing. The magnitude of the transient is directly proportional to the reverse-flow velocity when the valve closes (see Eq. 4.16). This velocity is controlled by the dynamics of the valve and the system. For a system where the flow reverses slowly, most valves will close before any significant flow reversal occurs, especially if the valve is spring-loaded. If flow reversal occurs rapidly, a relative high reverse velocity may occur before closure. For pumps in which air chambers are installed near the pump, rapid flow reversals occur. Simple check valves are not recommended for this application. A similar situation occurs when two or more pumps are working in parallel feeding one line and one pump is turned off.

For a given valve, the maximum reverse velocity is a function of the rate of flow reversal in the pipe. This information can only be obtained experimentally in special test facilities. Such information is scarce and available only on a few valves (44). To the author's knowledge, no such information is currently available on valves manufactured in the United States.

The rate of flow reversal will be different for each system. It can only be obtained by performing a transient analysis. Techniques used to complete such an analysis are discussed in subsequent chapters. Assume such an analysis indicates that the maximum reverse velocity would be about 0.7 m/s. If the pipe has a wave speed of 1100 m/s, the transient head rise would be about $(1100 \cdot 0.7/9.81) = 78$ m.

The reverse flow velocity and the associated transient head rise vary with valve type, size and the dynamics of the system. The same valve installed in different systems can behave differently. Transient analyses should be carried out for nonreturn valves to avoid transients. Unfortunately, little experimental information describing their closing behavior is available.

There are also instability problems associated with check valves which are usually caused by the valve being installed at a location where there is excessive turbulence or unstable flow. This causes the disk to flutter much like a flag in the wind and causes rapid wear, which can result in valve failure. This situation is aggravated if the valve disk operates in a partially open position. If the disk is not firmly seated, it can flutter and cause rapid wear.

## 4.6 AIR VALVES

For proper performance of a pipeline, a means of expelling air during filling, draining, and operation must be provided. Trapped air causes numerous hydraulic problems and should be avoided. When a pipeline is drained, air must be allowed to enter the line. The removal and admitting of air are normally handled with automatic air valves. Manually operated valves can be used, but they are not recommended because they can cause serious problems.

There are three types of automatic air valves: 1) air-vacuum valves; 2) air-release valves; and 3) combination valves. Figure 4.12 shows a simplified sketch of the first two types. The air-vacuum valve generally has a large orifice (1/2–36 in.) to allow large quantities of air out when filling or when draining a line. Air valves are mounted vertically on top of the pipe. They contain a float, which settles by gravity when the pipe is empty or the pressure is negative. As water enters the valve, the float rises and seals off the orifice. Once the line is pressurized, the valve cannot reopen to remove air that may subsequently accumulate. If the pressure becomes negative due to a transient or draining, the float drops and admits air into the line.

Air-release valves contain smaller (general 1/2 in. or smaller) orifices and are intended to release small quantities of air that accumulate after initial filling. The small orifice is controlled by a plunger activates by a float at the end of a lever arm. As air accumulates in the valve body, the float settles

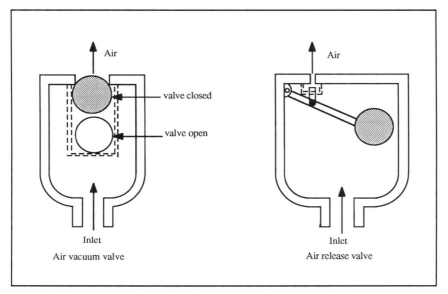

**Fig. 4.12**  Typical air-release valves.

and opens the orifice. As the air is expelled, the float rises and closes off the orifice.

The combination valve consists of an air-vccuum valve in parallel with an air release valve. The two can be separate or inside the same body. The choice of which valve or combination of the three types of valves to use depends on several factors: 1) how often the pipe is drained and filled; 2) how much air is entrained at the intake, in regions of low pressure or due to operation of vacuum valves; 3) the profile of the pipeline; and 4) the pipe's resistance to collapse due to external pressure and internal vacuum. When in doubt, the best choice is the combination valve.

Selecting locations to install air valves depends primarily on the pipe profile. Ideally, the pipe should be laid to grade with valves placed at all high points, or at about 1/4-mile intervals along the pipe if it has a constant slope or is level. When a pipe follows the terrain, it is desirable but not always feasible to place a valve at each high point. This situation should be avoided when possible.

The velocity of the liquid in the pipe is also important. If the velocity is adequate (above about 3 fps), trapped air can usually be flushed through the pipe to the air release valves. However, one must be careful about allowing large quantities of air under high pressure to accumulate and move through the pipe. Transients can be generated due to the movement of air, especially if it is improperly released from the pipe or if it passes through a control valve. More will be said about transients caused by trapped air in later chapters.

Air-vacuum and combination valves should be sized so that the air is expelled without pressurizing the pipe. Sizing charts are provided by manufacturers. When sizing an air valve to remove air during filling, the filling rate must also be considered the filling rate. A safe filling rate depends on the specific system. A conservative value frequently used is 1 fps. To accomplish this procedure safely, it may be advisable to provide a separate small filling valve or small auxiliary pump. One should never try to control the filling rate with the air valves.

For draining the pipe, air-vacuum valves must be installed. The number of valves depends on the number of pronounced high points in the pipe profile, the collapse pressure of the pipe, and the desired draining rate. The water can leave only as fast as air is admitted. Some consideration should be given to pipe rupture and subsequent rapid draining of the pipe. If the negative pressures caused by this situation can collapse the pipe, properly sized air valves should be installed. They should be sized so they can admit air flow rates equal to the maximum water flow caused by a full pipe break at the lowest point. If risk of pipeline collapse due to vacuum does not exist, air and vacuum valves may be installed to prevent water column separation.

One of the worse ways to remove air from a pressurized pipe is to open a large manual air valve after the pipe is pressurized. Consider a pipe with several high points, each with trapped air at high pressure. If a large manual air valve is opened, the air escapes rapidly because of its low density. The compressed air trapped at adjacent high points accelerates the liquid toward the open air valve at high velocity. When the air is all gone, the water cannot exit as fast, and a large velocity reduction can occur which can generate a severe transient. If air is to be released under high pressure, it must be done slowly.

One way to reduce the transient caused by rapid closure of an air valve is to use a hydraulically controlled air valve. The hydraulic dashpot allows the closure speed of the valve to be controlled.

## PROBLEMS

**4.1.** A 6-in. globe valve is tested in the laboratory to evaluate its flow coefficients. At a flow rate of 1200 gpm, the net pressure drop is 6.3 psi. Calculate $C_v$, $C_d$, $C_{d1}$, and $K_l$.

**4.2.** Water will be supplied to a small hydraulic turbine through a 408-mm pipe. It is desired to install a 305-mm butterfly valve to control the flow. The $C_d$ of the valve fully open is 0.75. Calculate the head drop across the valve when the flow rate is 0.23 m$^3$/s.

*Answer:*   0.39 m

**4.3.** Water is supplied between reservoirs through a 6-in. pipe. $\Sigma(K_l + fL/d) = 59$ when the control valve is fully open. For a head difference of 125 ft between the two reservoirs, calculate the flow rate. Use data for the butterfly valve shown in Fig. 4.3. Calculate the valve opening required to reduce the flow 50%.

*Answer*:  20%

**4.4.** A 12-in. butterfly valve has torque and flow coefficients similar to valve A shown in Figs. 4.7 and 4.8. If the flow rate is 25 cfs when the valve is half open, calculate the hydrodynamic torque.

*Answer*:  ~400 ft-lb

**4.5.** A 16-in. cone valve has an operator with a torque limit of 200 ft-lb. Find the maximum allowable velocity at 83° open (peak value). (Use data for 12 in. cone valve, Figs. 4.3 and 4.5.)

*Answer*:  18.4 fps

**4.6.** For example 4.1, case B, select a butterfly valve size so that at 50% open it has a head loss at least 15% of the total head loss available. Assume the pipe size is 12 in. in diameter.

*Answer*:  8 in.

**4.7.** If the 24-in. butterfly valve (style of valve A) in Example 4.2 were instantly closed from 40° open, calculate the head rise in the pipe. Assume $\mathbf{a} = 3100$ fps.

*Answer*:  1425 ft

**4.8.** Estimate the valve opening for Example 4.2 where it can be instantly closed and keep the maximum system head rise below 400 ft. Assume $\mathbf{a} = 3100$ fps.

*Answer*:   $\sim 12°$
(based on  $C_d = 0.0332$)

**4.9.** In Example 4.5, what would be the calculated flow through the bypass valve if: friction and minor losses were ignored, (b) $d2 = 1$ m., and $d3 = .626$ m.

*Answer*:  18.1 m$^3$/s

**4.10.** An 8-in. globe pressure-reducing valve is installed in a 12-in. line ($fL/d = 150$) between two reservoirs with a 175-ft difference in elevation. The valve will control between 20 and 60% open. Calculate the range of flows and $\Delta P_s$ that the valve will produce.

*Answer*:  60%: 5.62 cfs, 55.5 ft, Cd = 0.26
20%: 1.44 cfs, 167 ft, Cd = 0.04

# 5

# FUNDAMENTALS OF CAVITATION

Cavitation consists of rapid vaporization and condensation of a liquid. The process is somewhat analogous to boiling. For boiling, vapor cavities are formed from nuclei by increasing the temperature with the liquid at constant pressure. As the temperature increases, eventually vapor cavities are formed which rise to the free surface. As the cavities rise, they expand due to decreasing pressure and increased vaporization. At the surface, the vapor cavities explode, releasing the vapor to the atmosphere.

Cavitation normally occurs when liquid at constant temperature is subjected to vapor pressure either by a static or a dynamic means. If the local pressure somewhere in the fluid drops to or below vapor pressure and nuclei are present, vapor cavities can be formed. As long as the local pressure stays at vapor pressure and the cavity has reached a critical diameter, it will continue to grow rapidly. If the surrounding pressure is above vapor pressure, the bubble becomes unstable and collapses. The collapse can be violent and is accompanied by noise, vibrations and possible erosion damage to solid surfaces.

## 5.1 TYPES OF CAVITATION

There are two types of cavitation described in the literature: gaseous cavitation and vaporous cavitation. Gaseous cavitation occurs when there is either considerable free air suspended in the liquid or when the cavitation process is slow enough that the amount of air inside the vapor cavity is increased due to degassing from the liquid. The rate of growth and collapse of the cavity is slower for gaseous cavitation because of the presence of the free air. Consequently, the process is not as violent or damaging.

If there is little air in the liquid, so that the cavity consists almost exclusively of vapor, the growth and collapse rates and the pressures generated upon cavity collapse are extremely high and can cause severe damage. Called vaporous cavitation, this is the type of cavitation which is of concern to the practicing engineer. Gaseous cavitation will be discussed mainly as it relates to a means of suppressing vaporous cavitation by aeration of the fluid.

## 5.2 EFFECTS OF CAVITATION

There are a few beneficial effects of cavitation. For example, the high degree of turbulence associated with cavitation can be used to increase mixing, accelerate chemical reactions, and assist with homogenizing milk. Cavitation can also be used to clean surfaces with ultrasonic cleaning devices. However, as applied to hydraulic systems, the effects of cavitation are almost always detrimental. There are five basic problems created by cavitation: noise, vibrations, pressure fluctuations, erosion damage, and loss of efficiency or flow capacity.

The type and intensity of the noise depend on the item being considered and particularly on its size. For example, cavitation in a small valve is usually heard as a hissing or a light crackling sound. In a large valve, the noise can sound more like dynamite exploding. The sound varies significantly with valve design. Some large valves divide the flow into many small passages and produce a cavitation sound more like small valves.

The following description would be characteristic of the sound generated by a cone or butterfly valve about 12 in. (150 mm) in diameter operating in a gravity system where there is little background noise. In the earliest stages, cavitation is usually heard as light intermittent crackling sounds. These are just slightly louder than the basic turbulence noise generated by the flowing system. As the cavitation level increases, the intensity and frequency increase so that it is easy to hear the cavitation above the operating level of the system. When the valve is operating at a moderate level of cavitation, the sound resembles the noise that would be generated by gravel flowing through a pipeline. As the cavitation level approaches what would be called heavy cavitation, it consists of a continual loud roar and sometimes intermittent loud bursts of cavitation similar in sound to small explosions. At this level, it is almost impossible to converse near the valve; the sound level can exceed 100 db and can constitute a hearing hazard if exposure to such a level is for a long period of time.

The shock waves generated by collapsing cavities produce pressure fluctuations and cause the system to vibrate. As cavitation increases, the magnitude of the vibrations increases by several orders. Even for large valves tied down securely, the pipe and valve can move when operating at heavy levels of cavitation. Such vibrations can loosen bolts, cause fatigue of connections, loosen or break tie-downs, and lead to structural failure.

If the cavities collapse close to a solid boundary, erosion can occur. This is perhaps the most common cavitation problem. Many valves, pipes, pumps, turbines, etc., have been repaired or replaced due to excessive erosion caused by cavitation.

At advanced stages of cavitation, large vapor cavities form which can change the hydrodynamics of the flow through the system and reduce the efficiency of the device. Pumps produce less head, turbines generate less power, and valves no longer pass the predicted flow if they operate at advanced stages of cavitation. For example, the required net positive suction head for a pump represents the flow condition at which the cavitation level is severe enough to cause the efficiency of the pump to decrease between 1 and 3%. For a valve, onset of choking cavitation represents the condition at which the discharge coefficient of the valve decreases because of heavy cavitation.

## 5.3   ORIGIN OF CAVITATION

There are three fundamental requirements for cavitation to occur. First, there must be nuclei in the system, the presence of which serves as a basis for vaporization of the liquid. Second, the pressure somewhere in the liquid must drop, at least briefly, to or below vapor pressure. Third, the ambient pressure around the vapor cavity must be greater than vapor pressure in order for it to collapse.

### Role of Nuclei

The term nuclei is merely another word for gas bubbles or voids in the liquid. For either boiling or cavitation to occur, there must be nuclei present. If a liquid was completely deaerated and the container cleaned so that there were no contaminants, voids, or entrapped air either in the water or on the boundary, the liquid could sustain tension and would not boil at normal temperatures or cavitate when the pressure dropped to vapor pressure. Therefore, the existence of nuclei is one of the primary requirements for cavitation to occur. The primary sources of nuclei are from free air bubbles and air bubbles trapped in crevasses of suspended material or crevasses in the boundary.

The quantity and size of the nuclei depend on the history of the water. Water normally contains enough nuclei and contamination that cavitation will occur when the local pressure drops to vapor pressure. Only under controlled laboratory conditions can the nuclei content be reduced to the point where cavitation is significantly suppressed. If water is recirculated through a water tunnel or closed loop pumping system, the nuclei spectrum and air content are artificially changed and the condition at which onset of cavitation occurs can be altered.

Reducing the nuclei content as a means of reducing cavitation is only practical in the laboratory. On the other hand, it is easy to aerate the flow, causing an excessive amount of large nuclei as a means of suppressing cavitation.

Water removed at low levels from high head reservoirs may be low in air content and nuclei. Such water may suppress cavitation slightly. This would make laboratory tests with air-saturated water conservative. In contrast, if water is slightly supersaturated, either accidentally or intentionally, cavitation may occur slightly sooner, but it would likely be a gaseous type of cavitation and would be less damaging to the system.

## Sources of Low Pressure

The local pressure is the sum of the mean pressure, which is uniform over a certain region of the flow, and the dynamic pressure, which depends on fluid motion and especially on the formation and dissipation of eddies or vortices in turbulent shear zones. The mean pressure throughout a system varies due to changes of elevation, local and friction losses, and local accelerations due to a change of cross section of the flow. For large diameter tunnels, there can also be a variation of mean pressure from the top to the bottom. Such variations may be important in analyzing cavitation potential and should be properly accounted for in model studies.

Flow through a venturi provides a means of describing the influence of reduced mean pressure on cavitation. In a well-designed venturi, there is no flow separation and little turbulence. Consequently, for cavitation to occur in a venturi, the mean pressure at the throat must drop approximately to vapor pressure. However, there is always a boundary layer which generates turbulence. Cavitation will begin near the center of the boundary layer inside the tiny eddies, which have pressures at their center less than the mean pressure of the system.

When designing or analyzing a piping system, consideration should be given to the location of valves, pumps, or turbines and to the effect of mean pressure on cavitation. Since the chance of cavitation reduces as the mean pressure increases, it is desirable to place any device which might be subject to cavitation at a location of maximum pressure. This can be done, for example, by placing a valve at the lowest point in the pipeline, by lowering the pump relative to its suction piping, or by lowering a turbine relative to the downstream water surface elevation.

If an adequate pressure is not possible by adjusting the elevation of the installation, it may be desirable or necessary to increase the pressure by some other control device. For instance, if a valve is required to produce a large pressure drop and there is concern about cavitation, it may be desirable to consider placing an orifice or second valve downstream; this will create a moderate pressure drop and increase the pressure at the main control valve. For a pump, it may be desirable to increase the size of the suction piping or change the pipe configuration to reduce intake losses.

## Role of Eddies

Time-dependent pressures generated by turbulence or the formation and decay of eddies are an important part of the cavitation process. This time dependent pressure fluctuation is also the primary reason why cavitation is so difficult to predict and depends almost exclusively on experimental data. Mean pressures can be predicted by applying the energy equation and properly estimating any losses in the system. In contrast, the pressure inside separation regions, eddies, or vortices can only be estimated from empirical data.

To demonstrate the role of eddies or turbulence in the cavitation process, the flow characteristics of the sudden enlargement shown in Fig. 5.1 will be described. The flow in the approach pipe is at relatively low velocity and high pressure. As the flow approaches the orifice, the velocity increases and the pressure reduces. When the jet enters the downstream enlargement, an intense shear layer is created along the boundary between the jet and the surrounding separation region. The high velocity gradients in this shear layer create eddies, shown in the figure. The mean pressure in the separation region near the jet is about the same as the pressure inside the jet. However, the pressure inside the eddies is significantly less because of its high rotational speed. If a nucleus is entrapped in one of these eddies and the pressure inside drops to vapor pressure, it will begin to grow. If the pressure remains near vapor pressure long enough for the nucleus to reach a critical diameter, it then begins to grow rapidly by vaporization.

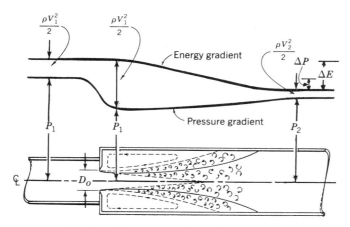

**Fig 5.1**  Flow pattern for submerged jet.

As the size of the vapor cavity increases, the strength of the eddy is rapidly destroyed, the rotational speed reduces, and the pressure is no longer vapor pressure. Since the ambient pressure is above vapor pressure, the cavity becomes unstable and collapses inward.

The time that a nucleus is subjected to the low pressure inside the eddy is important. If the time is so short that the bubble cannot reach its critical diameter, it will not become a cavitation event. Since the time of exposure increases as the system size increases, cavitation would be expected to be more severe in larger systems.

Compare the cavitation potential of a venturi to that of a submerged jet. Consider that both are installed in the same sized pipe and that the throat diameter of the venturi is the same as the contracted jet diameter for the submerged jet. If both are operating at the same upstream pressure and discharge, the submerged jet will be much more susceptible to cavitation because of the high degree of turbulence generated in the shear layer around the jet.

In some systems, it may be desirable to provide streamlining to avoid separation and cavitation. A typical example is the design of pump impellers. However, if energy-dissipation is required (as with a pressure-reducing valve), it may not be practical or possible to streamline to avoid separation. By so doing, the energy dissipation or pressure reducing capability of the device is reduced. Therefore, if one is required to produce both a large pressure drop and avoid cavitation, elimination of separation and turbulence is not the solution.

**Pressure Recovery**

The third phase of cavitation has already been alluded to: there must be a local pressure in the cavitation region greater than vapor pressure in order for the cavity to collapse. For systems such as a venturi, the high pressure required for collapse occurs as the vapor cavity moves from the throat region into the diffuser, where the pressure recovers with distance. There is also some pressure recovery as the tiny vortices in the boundary layer dissipate. For the orifice, the pressure recovery occurs primarily as a result of the eddies being dissipated and the cavity being subjected to the local pressure. There is also some pressure recovery associated with diffusion of the jet, as shown by the pressure gradient in Fig. 5.1.

## 5.4  DAMAGE MECHANISM

If the vapor cavities are transported to the solid boundary before they collapse, erosion damage can occur. Prior research has indicated that collapse must occur approximately one bubble diameter from the boundary in order to cause erosion damage. Since the bubbles are generally small, this indicates that only collapses very near or on the surface will cause erosion damage.

There are two suggested mechanisms by which damage occurs at solid boundaries. The first is the high-pressure shock waves generated by collapse of the cavities. These pressures have been estimated to be over $10^6$ psi, which is sufficient to damage any material. The other source of potential damage is due to what is called a microjet. When a bubble collapses near the boundary, the pressure distribution around the bubble is unsymmetrical due to the presence of the boundary. As the bubble collapses, the side of the bubble away from the wall attains a higher velocity and the bubble collapses inward, forming a jet shooting through the center of the bubble. This jet attains high velocities and creates a local pit when it impacts the wall.

Once a system reaches a point at which erosion damage occurs, damage increases rapidly as the velocity of the system is increased. Previous research has indicated that the rate of erosion damage varies with $V^n$, where $n$ varies, depending on the experiment, between 4 and 7 (33). This extreme dependence of damage rate on velocity suggests that one be conservative in selecting conditions corresponding to onset of erosion damage, since a slight increase in the velocity causes a large variation in the damage rate.

Another aspect of cavitation erosion damage is corrosion. Cavitation removes oxidized material from the surface, exposing fresh metal, which greatly accelerates corrosion. It is possible that a significant percentage of material removed by cavitation is due to corrosion. This concept has been used as a laboratory method of detecting cavitation damage (9).

Cavitation can also accelerate erosion damage caused by sediment in the liquid. The surface is weakened by the cavitation and is more easily eroded by the abrasive action of the sediment.


## 5.5    SUPPRESSING CAVITATION

Understanding cavitation and the damaged mechanism suggests several means of reducing or eliminating such damage. One method of eliminating damage is to restrict the operation of the system to a low level of cavitation such that the cavitation events do not contain much energy and collapse before they reach the boundary. Material in subsequent chapters will identify the limiting operating condition of various types of devices and will aid in designing systems that can operate cavitation free.

A second possible method for suppressing erosion damage is removing the boundary from the cavitation zone. One of the effective methods of dissipating energy while suppressing cavitation is the use of sudden enlargements. For example, if the diameter of the downstream pipe in Fig. 5.1 were increased, that system could operate free of cavitation damage at larger pressure differentials. One of the disadvantages of this approach is that the loss coefficient of the sudden enlargement increases as the expansion ratio increases. If large pressure drops are desirable, this may not cause a problem. However, if it is necessary to operate a system with low head loss, such a solution may not be acceptable.

A third means of controlling boundary erosion is by treating the surface with a material more resistant to cavitation erosion. It is common to line certain parts of valves, pumps, and turbines with stainless steel or other erosion resistant materials. Such a solution is usually expensive and may not be cost effective. This will be discussed further in Chapter 6.

A fourth suggested means of controlling cavitation is by injecting air into the separation regions. This supersaturates the region with air, greatly reducing the bulk modulus of the liquid. This cushions the collapse of the cavities by reducing the wave speed and reduces or eliminates the erosion damage. This concept is most useful if the air can be forced in by atmospheric pressure. If the local pressure is above atmospheric pressure, a source of compressed air would be required and the solution would be more expensive and less reliable, and a larger mass of air would be injected. For pumps, turbines, and some valves, this may not be practical since it is difficult to get the air to the cavitating region. One possible solution is to inject air upstream and allow it to diffuse through the flow. Situations in which it is easy to use air include orifices, certain valves, and spillways. Placing air slots on spillways is the most reliable means of protecting the concrete surface from damage. A crucial factor to consider when contemplating the use of injecting air is whether or not the system can tolerate it. Air in pipelines can result in severe transients, affect the accuracy of flow meters, reduce the capacity of the line by collecting at high points, and plug sand filters at water-treatment plants, etc.

A fifth method of suppressing cavitation is to dissipate the energy in stages. By locating multiple valves or orifices in series the level cavitation of each stage can be reduced.

## 5.6 CAVITATION PARAMETERS

Quantifying the flow conditions at a given level of cavitation requires a dimensionless similarity parameter. The form varies with the type of device being considered. It was discussed in Chapter 3 that cavitation for a pump can be quantified with the required net positive suction head (NPSHr) or a cavitation parameter equal to NPSHr/$H_p$.

For valves and other devices that create a head loss, two commonly used forms are

$$\sigma = (P_d + P_b - P_{va})/\Delta P = (P_d - P_{vg})/\Delta P \qquad (5.1)$$

$$k_c = \Delta P/(P_u - P_{vg}) \qquad (5.2)$$

In the equations, $P_d$ is the pressure measured about 10 diameters down-stream and projected back by adding the friction loss, $\Delta P$ the net pressure drop, $P_u$ the pressure just upstream, $P_{va}$ (Table 1.1) the absolute vapor pressures, and $P_b$ the barometric pressure. $P_{vg} = P_{va}\text{-}P_b$. Fig. 1.1 shows graphically the relationship between $P_{vg}$, $P_{va}$ and $P_b$.

The choice of Eqs. 5.1 or 5.2 is arbitrary. For installations where $P_d$ is known or is constant, Eq. 5.1 is more convenient. The two are related by the Equation.

$$k_c = 1/(\sigma + 1) \tag{5.3}$$

Most of the data in this book are given in terms of Eq. 5.1.

For cavitation caused by surface roughness, an isolated roughness, an offset in the boundary, or by any device for which it is not possible or convenient to evaluate $\Delta P$, another form of sigma is convenient:

$$\sigma_2 = 2(P_u - P_{vg})/\rho V^2 \tag{5.4}$$

in which $\rho$ is the fluid density and $V$ the mean velocity. This definition is related to the other forms by

$$\sigma_2 = (\sigma + 1)K_l \tag{5.5}$$

There are a variety of other forms possible. The reader is free to use any form he or she is familiar with and convert data into any other form.

## 5.7   EVALUATING CAVITATION LIMITS

An analytical solution to the problem of cavitation has not been found. Emphasis has, by necessity, been concentrated on laboratory experiments to explore the physics and develop empirical relationships for predicting vari-ous levels of cavitation. The purpose of this section is to give some understanding of how cavitation is detected and how the data are analyzed. Such knowledge aids in understanding the cavitation process, which is helpful in applying the various cavitation limits and scale effects which are used for analysis and design.

Since cavitation causes noise, pressure fluctuations, vibrations, pitting, and loss of efficiency, any sensing device that is sensitive to one of these can be used to detect cavitation. Several electronic instruments, including sound level meters, pressure transducers, accelerometers, and hydrophones, have been used in the laboratory to measure cavitation.

The accelerometer has been the instrument most widely used in the laboratory by the author to detect the onset of cavitation (69). It is simple to use, rugged, portable, and has a wide frequency response. It is sensitive to the lightest cavitation and can measure the heaviest levels possible.

The manner in which the output from the various cavitation sensing devices is processed can vary with the type of testing. For example, to determine onset of cavitation in a pump, it is most helpful to filter out low frequency disturbances, since the background noise level generated by the pump can make cavitation detection difficult. For such an application, one could use a hydrophone, a high-frequency response pressure transducer, or the accelerometer with proper filtering. For low flow noise systems, it is helpful but not necessary to filter out the low frequencies.

Other methods of detecting cavitation are visually and aurally. If a transparent section can be installed, cavitation events can be seen either by the naked eye or with stroboscopic lighting. Even if it is not possible to visualize the flow, cavitation can usually be heard by ear except in systems where there is an extremely high background noise level. A trained observer can usually be quite precise in detecting early stages of cavitation either visually or aurally. In all cavitation studies, it is best to record the observer's observations to supplement the electronic measurements of cavitation. The procedure for evaluating incipient (initial onset of cavitation) and critical (light steady noise) for valves and orifices is discussed in Section 6.1.

Choking cavitation can sometimes be approximated by using electronic instruments that measure the cavitation intensity. However, it is normally evaluated by determining the effect of choking on the output of a pump or the discharge characteristics of a valve. For a valve, the procedure usually used is to set the valve at the desired valve opening, establish a $P_u$ and flow so the valve is not cavitating, and measure flow and pressure drop across the valve. This process is repeated at increased flow rates while maintaining $P_u$ until vapor pressure exists at the downstream piezometer tap. When this occurs, the flow is fully choked.

A condition of choking is demonstrated in Fig. 5.2 for an orifice. The upper line (#1) represents a normal hydraulic grade line before the onset of choking. Minimum pressure occurs at the vena contracta and normal

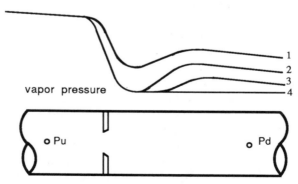

**Fig 5.2**   Hydraulic grade line (HGL) for choking cavitation.

pressure is recovered. Line #2 represents the condition where the flow rate has increased so that the minimum pressure at the vena contracta has just dropped to vapor pressure. For an orifice, this condition can be predicted with the Bernoulli equation and the contraction coefficient. When this occurs, the orifice is passing its maximum discharge $Q_{ch}$ at a given $P_u$. Up to this point, the discharge coefficient remains constant and equal to its value for noncavitating flow.

If the pressure in the downstream pipe is reduced, the pressure drops (line #3) and the vapor cavity lengthens, but the flow remains constant. Since $P_d$ reduces and $Q_{ch}$ remains constant, the discharge coefficient reduces. As the downstream pressure is further reduced, eventually the vapor cavity extends beyond $P_d$ and $P_d = P_{vg}$. This is sometimes referred to as supercavitation. If the pipe were transparent, one would see what appears to be a free discharge jet. The difference is that the jet is surrounded by a vapor cavity instead of air. At some downstream location where the pressure and momentum forces balance, the pipe again flows full and a normal pressure gradient is established.

One of the interesting aspects of supercavitation is that damage may not occur at the valve or orifice operating in supercavitation. However, at the downstream location where the cavity ends, severe damage will occur. Care must be taken because the worst condition at which to operate a valve, orifice, or pump is near the point of onset of choking, since erosion damage, noise, and vibration are maximum just before a valve chokes.

For evaluating incipient damage, soft aluminum inserts are cut, polished, inspected for surface imperfections, and installed flush with the inside surface of the device being tested. The size and location of the specimen depend on the size and configuration of the device being tested. The system is then operated for a given time, the specimens removed, and the cavitation pits counted. Incipient damage is defined as a pitting rate of 1 pit/in.$^2$/min. To date, damage data have been obtained for orifice plates plus a few valves and pumps.

## 5.8 SCALING AND SCALE EFFECTS

Virtually all cavitation data are obtained experimentally, often with reduced size models operated at reduced pressure and velocity. If complete similitude is not achieved geometrically, dynamically, and in terms of the fluid properties, the experimental data may not accurately represent the performance of the prototype. Such discrepancy between model and prototype is referred to as a scale effect. This must be distinguished from "scaling" data. "Scaling" refers to extrapolating data from one condition to another assuming complete similitude. For example, assume tests on a model valve at some downstream pressure indicated its critical cavitation index $\sigma_c = 2.1$. Ignoring scale effects, the prototype valve would have the same $\sigma_c$ at any pressure. One could therefore "scale" the experimental data and predict the $\Delta P$ across the prototype valve at critical cavitation for any $P_d$ using the sigma in Eq. 5.1. Scaling data from a model pump to a prototype pump simply involves assuming that the NPSHr is the same for both.

Unfortunately, cavitation is a classical example for which the techniques of dimensional analysis have not produced a sigma that is free of scale effects for all conditions. The difficulty is caused by the numerous variables that affect the process. The primary variables that influence inception and subsequent levels of cavitation and are included in the derivation of sigma are 1) the critical pressure required to form a vapor cavity (usually assumed to be the liquid vapor pressure), 2) absolute pressure, 3) velocity or pressure drop, and 4) boundary geometry.

Some of the variables that are not included in the formulation of the cavitation parameter are factors influencing vaporization of the liquid, such as surface tension, air content, and water quality. There are also factors that influence the growth and collapse of the cavities, such as turbulence scaling and residence time of the bubble in the low-pressure zone, which are, in turn, affected by velocity and the size of the system.

If there are significant differences in the fluid properties between the fluid used in the laboratory and the prototype fluid, some errors can result. This is particularly a problem with cavitation tests in recirculating water tunnels. Continual recirculation alters the air content and nuclei spectrum of the water. Even when artificial means are used to control the nuclei, significant effects on onset of cavitation have been observed (23, 24).

Experimental data for valves (Chapter 6) show significant scale effects due to increased velocity, pressure, and size. The data show that $\sigma_i$ (incipient) and $\sigma_c$ (critical) for valves and some other devices increase with Reynolds number Re due to changes in velocity and size. The Reynolds number scale effects due to velocity changes may be related to turbulence scaling. For certain geometries, the magnitude of the fluctuating pressures responsible for cavitation apparently do not scale proportional to $V^2$. Another cause may be the assumption of vapor pressure as the critical pressure.

The Reynolds number scale effect caused by changing the size is most likely caused by increased residence time of the nuclei at low pressures. For a larger system, more of the nuclei can grow to a critical diameter because they are subjected to low pressures for a longer time due to the lower frequencies of the pressure fluctuations in larger systems and the longer residence time in the shear zone. As size increases, the nuclei are also influenced more by the higher frequency fluctuations because they are subjected to more cycles.

Cavitation data for orifices and valves show that Re is not a parameter that produces similitude for all cavitation conditions. By similitude, it is meant that if two systems were operating at the same Re, they should produce the same level of cavitation. The data in Fig. 5.3 show the variation

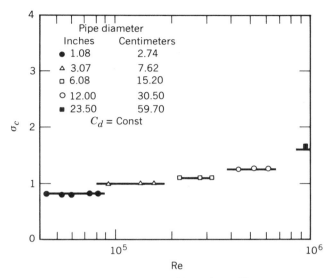

**Fig 5.3** Size scale effects for orifices.

of $\sigma_c$ with Re for orifices in different sizes of pipes for a constant ratio of orifice diameter to pipe diameter. It is observed from the figure that for a constant pipe size, $\sigma_c$ does not vary over a range of Re. In this case, Re changes by increasing the pipe velocity while varying the upstream pressure and the flow such that the cavitation intensity is constant.

If, however, the velocity and cavitation intensity are held constant and Re is increased by increasing the pipe diameter, the value of $\sigma_c$ increases. For certain values of Re, there are two values of $\sigma_c$. It is therefore apparent that there is an effect on $\sigma_c$ associated with changes in the size of the pipe which are not properly accounted for with Re. Additional information on scale effects for valves and orifices is presented in Chapters 6 and 7.

## PROBLEMS

**5.1.** Name the three basic requirements for cavitation to occur.

**5.2.** List the results of cavitation.

**5.3.** Describe the two mechanisms of cavitation damage.

**5.4.** How does free air influence the growth and collapse of cavities?

**5.5.** What are nuclei and where do they come from?

**5.6.** Under what conditions can water sustain tension? (Pressure less than vapor pressure.)

**5.7.** List some methods of controlling cavitation erosion.

**5.8.** An orifice is to be used in a system where its sigma (Eq. 5.1) will be 1.12. Tests are to be conducted in the lab to measure the cavitation intensity. Assume the tests will be conducted at $P_u = 200$ kPa, and $P_{vg} = -75$ kPa. What is the required $P_d$ to produce the same $\sigma$. (Ignore line loss.)

*Answer:*   $P_d = 70.3$ kPa

**5.9.** Assume the following information was obtained in the lab during a cavitation test on an orifice: $C_d = 0.10$, $P_u = 620$ kPa, $P_{vg} = -84$ kPa, $V_c = 2.69$ m/s. Calculate $\sigma$ (Eq. 5.1).

*Answer:*   $\sigma = 0.97$

**5.10.** A 6-in model of a 24-in. valve was tested in the laboratory and at a valve opening of 60° the following model data were found: $C_d = 0.57$, $P_u = 59.5$ psi, $V_c = 27.5$ fps (at critical cavitation). Calculate the discharge at critical cavitation for the 24-in. valve operating at $P_u = 95.4$ psi; ignore scale effects and assume $P_{vg} = -11.6$ psi (for both valves).

*Answer:*   $Q = 107$ cfs

**5.11.** A device having a $C_d$ of 0.34 for noncavitating flow is operating at $P_u = 650$ kPa and $P_d = 150$ kPa at a pipe velocity of 10 m/s. Is the device operation at choked cavitation? (Ignore line loss.)

*Answer:*   yes

# 6

# CAVITATION DATA FOR VALVES

Cavitation is frequently an important consideration in the design and operation of valve installations. It is necessary to determine if cavitation will exist, evaluate its intensity, and estimate its effect on the system and environment. Determining the existence and intensity of cavitation depends on the availability and proper application of experimental data. Cavitation varies with valve type, size, operating pressure, and details regarding the installation. Improper use of laboratory data can result in significant errors. Another aspect of satisfactory design is establishing an acceptable cavitation limit for a given system. One must consider the effects that the cavitation will have on the system or environment: noise, vibration, erosion damage, or a decrease in performance, for example. The selected limit may vary from requiring no cavitation to allowing fully choked flow.

This chapter discusses application of experimental data for predicting the cavitation intensity of valves. Items covered include:

1. A discussion of cavitation design criteria.
2. Laboratory data for valves.
3. Scale effects associated with each cavitation limit.
4. Examples to demonstrate the application of the data.

## 6.1 CAVITATION DESIGN CRITERIA

Selecting design criteria is partially controlled by the availability of experimental cavitation data. One would not select a design limit for which there was no available data or at least an established method for its evaluation. Choosing a cavitation limit depends on several factors related to the operating requirements, expected life, location of the device, details of the design, and economics. For example, if a valve is required for infrequent operation, such as for filling a pipeline or for pressure relief, the valve might be designed to operate with heavy cavitation. If the valve is for continual use, heavy cavitation would be avoided.

The location of the valve is an important factor, since noise and vibration caused by cavitation can be objectionable. If it were in a residential area, a light-to-moderate level of cavitation would be the maximum allowable; in a remote area, this perhaps would not be a design criteria.

Details of the valve and piping can influence the design limit. In the case of an orifice or a needle valve discharging into a sudden enlargement, advanced stages of cavitation can exist before significant erosion damage occurs on the expansion wall. The larger the expansion ratio, the less chance of cavitation damage. In this case, noise, vibrations, or loss of valve efficiency may be more important than erosion damage.

These and many other factors dictate the final choice of the cavitation limit. Since the cavitation level that can be tolerated varies considerably, several limits must be evaluated to give the designer adequate flexibility. Four limits will be discussed: incipient, critical, incipient damage, and choking.

**Incipient and Critical**

Incipient is a conservative design limit and is suggested for use only when no noise or other disturbances can be tolerated. Since it is so conservative, it has been evaluated only for a few valves. The next-higher cavitation level is critical. The cavitation at critical would not be objectionable and would not decrease the valve life. For most applications, the critical condition is recommended for what might be termed "cavitation-free operation." Considerable data has been obtained for critical cavitation.

The values of $\sigma$ at incipient $\sigma_i$ and critical $\sigma_c$ can usually be evaluated by plotting the accelerometer output versus $\sigma$ on log–log coordinates, as shown in Fig. 6.1 for a butterfly valve. Region a–b corresponds to no cavitation. The variation in accelerometer output with sigma in this region is due strictly to increased flow noise. Care must be taken in this region to ensure that the control valves or other components in the system are not cavitating or causing excessive turbulent noise. Region b–c covers incipient to light cavitation and c–d from light to the maximum intensity. Incipient and critical cavitation are determined by the intersection of straight-line portions of the figure. This method has been used on numerous valves and orifices and found to give repeatable results provided care is used to avoid extraneous disturbances during testing. A few valves tested have significantly different cavitation characteristics and produce different accelerometer curves.

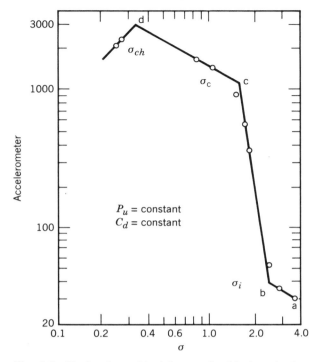

**Fig. 6.1**   Evaluation of incipient and critical cavitation.

There is usually a sudden drop in the cavitation intensity as the valve begins to choke (point d). This is because as the length of the vapor cavity downstream from the valve increases, the collapse of the cavity occurs farther from the valve. When the valve is fully choked (sometimes referred to as flashing or supercavitation), the vapor pressure zone may extend for hundreds of pipe diameters (69).

### Incipient Damage

Little information is available identifying the flow conditions corresponding to onset of erosion. This is mainly due to the experimental difficulties involved in obtaining such a limit. The limit corresponds to conditions where pitting is first detected on a soft aluminum specimen placed at the boundary. Soft aluminum is used as the test material to minimize testing time. The intensity of the noise and vibrations at incipient damage varies, depending on the test device. For an orifice with a small orifice-to-pipe diameter ratio, and for certain valves, fairly heavy cavitation is required before pitting occurs at the wall. Once pitting begins, only slight increases in the system velocity, or a small drop in ambient pressure, can rapidly accelerate the damage. Experiments have shown that the rate of weight loss is proportional to velocity raised to a power: damage is proportional to $V^n$. The exponent $n$ has been evaluated experimentally to be as large as 7 (33, 59, 60). Operating a valve beyond incipient damage, therefore, subjects it to the possibility of rapid erosion. Incipient damage data for butterfly, cone, and globe valves will be presented.

Incipient damage $\sigma_{id}$ is evaluated by plotting the experimentally determined pitting rate versus pipe velocity or sigma. Figure 6.2 shows such data for an orifice. Incipient damage is defined as 1 pit/in.$^2$/min on the soft aluminum. For tests on some valves, it is sometimes not possible to count individual pits. For such cases, it is necessary to observe when pitting first occurs by changing the flow conditions in small increments and observing the damage specimens. The steepness of the lines shows how rapidly damage increases as the velocity is increased. Figure 6.2 shows that for an increase of about 20% in the velocity, the pitting rate increases about 1000 times. It is therefore important to be conservative in estimating $\sigma_{id}$ and not to operate at $\sigma < \sigma_{id}$.

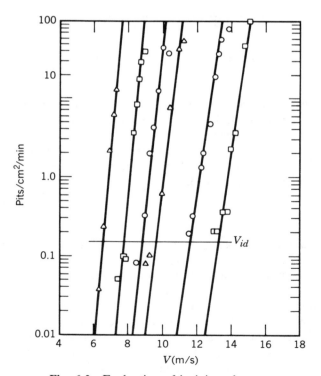

**Fig. 6.2**   Evaluation of incipient damage.

## Onset of Choking

Using choking cavitation as a design point is often appropriate for valves that only operate for short periods of time, such as a pressure-relief valve. Since this type of valve generally operates infrequently, erosion damage may not be the deciding factor. When a system is designed to tolerate heavy vibration and if noise is not a consideration, a valve can operate choked. As the flow increases through a valve or similar device, the mean pressure just downstream eventually drops to vapor pressure and the valve chokes. The term signifies that for a given upstream pressure, the valve is passing its maximum discharge. Near choking, the flow for certain valves is unstable. Only a slight change in the downstream pressure can cause a supercavitation condition in which vapor pressure extends for many diameters and the cavity collapse occurs far downstream. Under these conditions, some valves can operate almost damage-free. This large cavity can also disappear with only a slight pressure change downstream, causing a transient pressure rise.

The normal method of evaluating onset of choking does not require a measure of the cavitation intensity. The procedure is demonstrated in Fig. 6.3. In the absence of cavitation, and up to heavy cavitation, the discharge (or velocity) is proportional to the square root of the pressure drop. This is reflected in the equations for $C_d$, $C_v$, and $K_l$ (Eqs. 4.1–4.4). Up to the point of heavy cavitation, the value of the coefficients are constant, as reflected by the straight portion of the lines in Fig. 6.3 having a slope of 1:2. As the valve begins to choke, the slope of the line decreases and the line becomes horizontal as supercavitation is reached. The abruptness of the change in

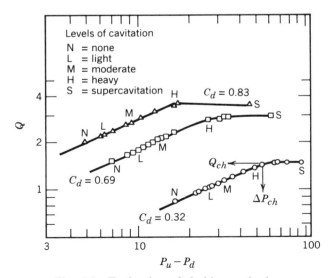

**Fig. 6.3**  Evaluation of choking cavitation.

slope is dependent on the type of valve and on the location of the $P_d$ measurement. The discharge $Q_{ch}$ and pressure drop $\Delta P_{ch}$ at choking are determined by intersecting the two straight lines, as shown in the figure. The value of $\Delta P_{ch}$ can also be calculated using the measured $Q_{ch}$ and the value of $C_v$, $K_l$ or $C_d$ for noncavitating flow. $\sigma_{ch}$ is calculated from Eq. 5.1 using

$$\sigma_{ch} = \frac{(P_u - \Delta P_{ch} - P_{vg})}{\Delta P_{ch}} \qquad (6.1)$$

The intensity of cavitation and the corresponding noise vibration and erosion damage at the valve are at their maximum just before the valve chokes. However, if the valve moves into supercavitation the situation is quite different. For valves such as butterfly and cone valves, most of the collapse occurs remote from the valve and little damage is likely to occur at the valve. However, farther downstream, where collapse occurs, serious problems may be encountered. For valves such as globe valves, some of the cavitation will collapse inside the valve body even with supercavitation and will cause erosion damage.

Some of the new high head loss valves that use the principle of multiple losses in series do not fully choke. Consider a system with several orifices in series designed so they do not cavitate under normal conditions. If the downstream pressure is dropped to vapor pressure, the last orifice may choke, but the others do not. This produces very little change in the valves discharge coefficient $C_d$ and appears as if the system is not really choked.

**Noise Level**

The noise generated by cavitation can be objectionable. Larger valves can easily generate over 100 db of noise when cavitating heavily. The noise is also influenced by the piping. A gradual expansion acts like a horn and can amplify the sound generated by the valve. Information is available on cavitation noise to provide design information in Reference 45.

## 6.2 EXPERIMENTAL DATA

Once the limiting cavitation condition has been established, the next step is to obtain experimental data on the valves being considered. In general, there are three sources of data: 1) tests on the actual valve conducted with conditions identical to those of the system; 2) tests on a model valve; or 3) data from the literature. The first source obviously gives the best information since it generally avoids all scale effects. The second source is reliable provided the data are properly adjusted for scale effects. The third source can provide data that will give a reasonable estimate if two valves are geometrically similar and if proper care is used in selecting and applying the data. This chapter provides experimental data on several common types of valves and demonstrates how to adjust for scale effects.

One of the frustrations of using cavitation data available in the literature is the apparent discrepancy between the values different researchers obtain for the same limit. This occurs because of several problems. First, since cavitation is a random process, defining and consistently reproducing a value for the cavitation limit is difficult. This can result in variation in the data even from the same research. This can be minimized by refining the method of selecting the limit. The methods of selecting cavitation limits described in this chapter eliminate much of the subjective judgment.

A second difficulty is in different definitions of the same cavitation limit. For example, one researcher (57) defined incipient cavitation as the point at which the discharge coefficient of the valve begins to drop off due to heavy cavitation. He uses a definition of the cavitation index

$$k_c = \frac{(P_u - P_d)}{(P_u - P_{vg})} = \frac{1}{\sigma + 1} \tag{6.2}$$

Defining $k_c$ as the point where $C_d$ drops off and calling it onset of cavitation have caused much confusion. The point at which the efficiency of the valve begins to decrease is near choking and actually corresponds to flow conditions near the maximum noise, vibration, and erosion potential for valves. It is similar to the required NPSH for pumps.

A third source of apparent scatter in what would at first appear to be similar data is due to scale effects. Much of the apparent scatter can be eliminated by properly making adjustments for scale effects. A fourth source of data variation is differences in valve detail for valves of the same general type. Certain geometry changes, such as the seat design, can measurably affect the onset of cavitation, whereas other minor variations do not. Other examples of significant differences are ball or cone valves, some of which have a solid or skirted plug and others of which have a fabricated or skeleton-type plug where flow can go around the plug as well as through it. Another example is a partial ball versus a complete ball. Ball valves constructed with a partial ball are made such that there is only one port where throttling occurs. These differences can cause substantial variations in the cavitation limits for valves bearing the same name. Care must be exercised when comparing valves of similar types. These principles are demonstrated with the experimental data presented subsequently.

### Incipient Cavitation Data

Since the incipient cavitation point represents a very conservative design limit it has been evaluated only for certain valves. Sample data are shown in Fig. 6.4 for a 6-in. butterfly valve and in Fig. 6.5 for a 6-in. cone valve. The cavitation index $\sigma$ and the discharge coefficient $C_d$ are defined by Eqs. 5.1

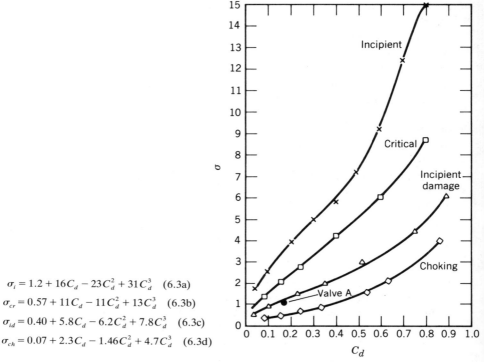

$$\sigma_i = 1.2 + 16C_d - 23C_d^2 + 31C_d^3 \quad (6.3a)$$

$$\sigma_{cr} = 0.57 + 11C_d - 11C_d^2 + 13C_d^3 \quad (6.3b)$$

$$\sigma_{id} = 0.40 + 5.8C_d - 6.2C_d^2 + 7.8C_d^3 \quad (6.3c)$$

$$\sigma_{ch} = 0.07 + 2.3C_d - 1.46C_d^2 + 4.7C_d^3 \quad (6.3d)$$

**Fig. 6.4** Cavitation data for a 6-in. butterfly valve ($P_{uo} = 70$ psi, $P_{vgo} = -12$ psi).

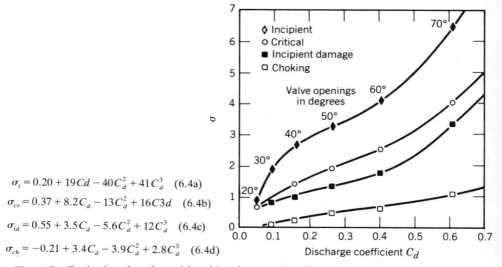

$$\sigma_i = 0.20 + 19Cd - 40C_d^2 + 41C_d^3 \quad (6.4a)$$

$$\sigma_{cr} = 0.37 + 8.2C_d - 13C_d^2 + 16C3d \quad (6.4b)$$

$$\sigma_{id} = 0.55 + 3.5C_d - 5.6C_d^2 + 12C_d^3 \quad (6.4c)$$

$$\sigma_{ch} = -0.21 + 3.4C_d - 3.9C_d^2 + 2.8C_d^3 \quad (6.4d)$$

**Fig. 6.5** Cavitation data for a 6-in. skirted cone valve ($P_{uo} = 40$ psi, $P_{vgo} = -12$ psi).

**140**

and 4.4. This limit represents an average cavitation level of about 1 event per second. It is primarily of academic interest because it is too restrictive for a practical design limit.

### Critical Cavitation Data

The greatest amount of cavitation data available for valves is for critical cavitation. This limit is referred to as incipient by some researchers (83) but their incipient point corresponds more closely to $\sigma_c$ as defined herein than it does to $\sigma_i$. Critical cavitation data have been obtained by the author on numerous types and sizes of butterfly valves. Each valve varies slightly. The data in Figs. 6.4–6.7 are representative and can be used for a first estimate. It is suggested that cavitation be obtained from the manufacturer if it is available. This applies to all cavitation data presented in this chapter.

Critical cavitation data for butterfly, cone, and ball valves are shown in Figs. 6.4–6.6. The cone valves have tapered conical plugs. When fully open, the flow passage has the same area as the pipe. The skeleton cone valve is fabricated such that at partial valve openings a considerable amount of water can flow around the plug, with the rest going through the main port. The skirted cone valve, like a solid ball valve allows water to flow only through the main port except for a small amount of leakage around the plug. The skeleton cone valve is similar to the skeleton ball valve in that

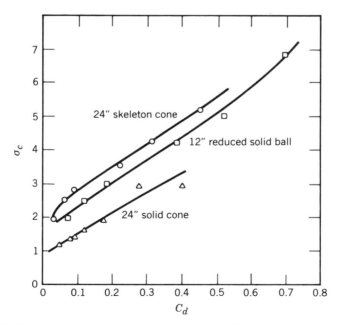

**Fig. 6.6**  Critical cavitation for ball and cone valves ($P_{uo} = 75$ psi, $P_{vgo} = -12$ psi).

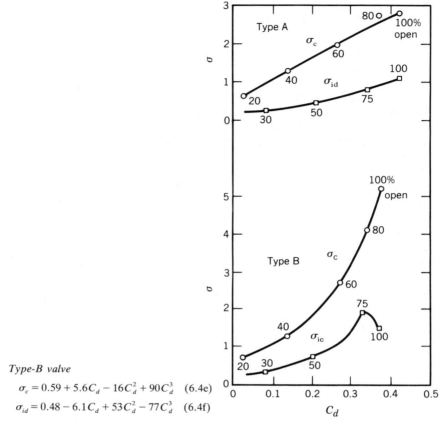

*Type-B valve*

$$\sigma_c = 0.59 + 5.6C_d - 16C_d^2 + 90C_d^3 \quad (6.4e)$$

$$\sigma_{id} = 0.48 - 6.1C_d + 53C_d^2 - 77C_d^3 \quad (6.4f)$$

**Fig. 6.7** Critical and incipient damage for 6 in. globe valves ($P_{uo} = 60$ psi, $P_{vgo} = -12$ psi).

water can flow around the outside of the plug. The valve in Fig. 6.5 is for a skirted cone valve.

Critical cavitation data for two 6-in. globe valves are shown in Fig. 6.7. Notice the large differences in $\sigma_c$ at the same value of $C_d$ for the two valves. This is why one should obtain data for each valve from the manufacturer. Unfortunately, not all manufacturers have such data.

For using the cavitation data in spreadsheet applications or in other computer programs, it is helpful to fit equations to the data shown in Figs. 6.4–6.7. Polynomial equations are listed at the left of Figs. 6.4, 6.5 and 6.7. These equations are used in example problems.

**Incipient Damage Data**

Only recently (15, 40, 47, 74) have there been attempts to determine the onset of cavitation damage for valves; thereby establishing criteria that can be used to develop designs in which damage will not occur even though cavitation is present. Limited data are available for butterfly, cone and globe valves (Figs. 6.4, 6.5 and 6.7).

Figures 6.4 and 6.5 summarizes cavitation limits of incipient, critical, incipient damage, and choking cavitation for a 6-in. (152-mm) butterfly valve (15) and a 6-in. cone valve (40). Since pressure scale effects exist for incipient, critical, and incipient damage, the data in the figure are only valid for the upstream pressure listed on the figures. Scaling the data to any other size and pressure requires scale effects adjustments. For choking cavitation, the $\sigma$ values shown in Fig. 6.4 are valid for any pressure or size.

Incipient damage data for two 6-in. globe valves are shown in Fig. 6.7. Due to the complicated geometry inside globe valves, there is some uncertainty as to whether the study actually located the zone where damage first occurs. The only way to have refined the data more would have been to cover the entire interior of the valve with the aluminum specimens, which was not practical. It is therefore suggested that some conservatism be used in applying the data presented in this section to account for this degree of uncertainty.

The differences in the damage characteristics of the two globe valve styles point out that the value of the cavitation limits is strongly dependent on body style. Caution should therefore be exercised if these data are used to estimate damage conditions for other valves.

### Choking Cavitation Data

Data for choking cavitation for cone, globe, gate, ball, and butterfly valves are shown in Figs. 6.4–6.8. Data for ball, globe, and butterfly valves in Fig. 6.8 are not for the same valves as in Figs. 6.4–6.7. Note that the sets of data for the two butterfly valves are similar. In contrast, the data for the two globe valves in Fig. 6.7 are significantly different. Again, it is stressed that the cavitation data included in the text for the various types of valves is for general information only. Specific information for the actual valve being considered for a design should be used.

The values of $C_d$ used in all the figures are for nonchoking conditions. When the valve chokes, $C_d$ reduces significantly so it does not have a unique value at a given valve opening. When calculating discharge at choking, $C_d$ cannot be used with the measured $P_u - P_d$. (This is demonstrated in Example 6.4.)

A parameter used by the Instrument Society of America (26) to designate choking cavitation is defined as:

$$F_l = [\Delta P/(P_u - P_{vg})]^{0.5} = k_c^{0.5} \qquad (6.5)$$

The parameter $F_l$ is referred to as the pressure-recovery factor. The significance of this is that at choking the valve operates as if it were free-discharging (but into a vacuum). Consequently, there is not the normal

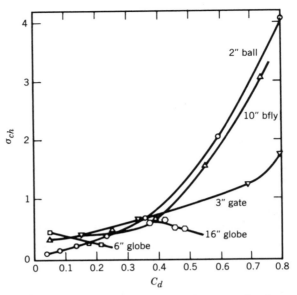

**Fig. 6.8**  Choking cavitation data for several valves.

pressure recovery downstream and the $\Delta P_{ch}$ is considerably larger than the net $\Delta P$ for normal pipe flow at the same discharge. This causes the $C_d$ and $C_v$ values to reduce significantly.

There are several ways to determine if a valve is choking. Any of the following tests can be used:

1.  $\sigma_{\text{system}} \leq \sigma_{ch}$    2.  $C_{d(\text{calc})} \leq C_{d\,(\text{no cavitation})}$    3.  $P_d = P_{vg}$
4.  $\Delta P \geq \Delta P_{ch}$    5.  $Q_{(\text{calc})} \geq Q_{ch}$

## 6.3   SCALE EFFECTS

For a review of scaling and scale effects, see Section 5.8. Cavitation scale effects are such that a large valve operating at a high pressure will cavitate worse than a small valve operating at low pressure. Some effort has been devoted to evaluate the cavitation characteristics of valves, orifices, elbows, etc., to establish values of $\sigma$ corresponding to the various levels of cavitation and to evaluate scale effects. One of the earliest publications on valves was by Yanshin in 1965 (88). A review of cavitation research on valves between 1957 and 1967 by Tullis and Marschner (70) provided values of $\sigma$ at different cavitation intensities for a variety of valves. Little information on scale effects was then available. Since 1967, considerably more information has become available. Much of this information has been accumulated by the author and his students (8, 10, 19, 45, 47, 53, 59, 60, 62, 63, 67, 71, 73, 74, 82), by the Metropolitan Water District of Southern California (83) and others (5–7, 49–52, 80). The impact of this research has been the evaluation of $\sigma$ corresponding to different levels of cavitation for a variety of types and sizes of devices plus the quantitative evaluation of scale effects associated with changes in the size and variations in operating pressure. For valves, onset of cavitation has been found to increase with diameter and with increased pressure (19, 62). The purpose of this section is to explain scale effects, quantify them, and demonstrate the proper application of ex-perimental cavitation data by identifying how to adjust for scale effects associated with variations in size and operating pressure. Such adjustments are necessary because failure to correct for these scale effects when using relatively small models can result in serious cavitation problems in the prototype.

It is important to note that when making scale effect adjustments, it is necessary to adjust the experimental sigma data and not the system sigma value. Adjusting the system sigma will give incorrect results.

### Pressure Scale Effects

For critical cavitation, Fig. 6.9 contains experimental data that show the variation of $\sigma_c$ with pressure for a 12-in. (305-mm) butterfly valve tested at several openings. The straight lines of approximately constant slope on the log–log plot suggest that pressure scale effects (PSE) can be made with the following equations:

$$\sigma_c = \text{PSE}\,\sigma_{co} \tag{6.6}$$

$$\text{PSE} = \left[ \frac{(P_d - P_{vg})}{(P_{do} - P_{vgo})} \right]^X \tag{6.7a}$$

or

$$\text{PSE} = \left[ \frac{(P_u - P_{vg})}{(P_{uo} - P_{vgo})} \right]^X \tag{6.7b}$$

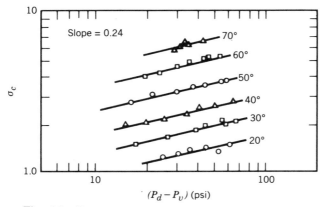

**Fig. 6.9**  Pressure scale effects for a butterfly valve.

in which $\sigma_{co}$ is an experimentally determined critical cavitation index evaluated at some reference test pressure, $P_{do}$ and $P_{uo}$, $P_{vgo}$ the experimental gauge vapor pressure, and the exponent $X$ the slope evaluated from experimental data similar to the data shown in Fig. 6.9. $P_u$, $P_d$, and $P_{vg}$ are the pressures at which $\sigma_c$ is desired. When there are no pressure scale effects, $X = 0$.

Incipient cavitation conditions can be scaled with Eqs. 6.6 and 6.7 by substituting $\sigma_i$ and $\sigma_c$. Equations 6.7 and 6.8 are not completely mathematically compatible. Slightly different values of $\sigma_c$ are obtained by the two equations. However, for the normal range of pressures, the differences are small and are about the same as the uncertainty in the experimental determination of $\sigma_{co}$.

Applying Eq. 6.6 requires 1) experimental data on a similar valve at a known pressure for the required valve openings (such as the data in Figs. 6.4–6.7), and 2) values for the exponent $X$. Values of $X$ for several valves are listed in Table 6.1. The seven butterfly valves listed in Table 6.1 represent extremes in body and disk design. The average value of $X$ for these valves is 0.28. The maximum deviation of $X$ from the mean value is about 15%. This amount of deviation generally causes an insignificant variation in the values predicted by Eq. 6.6 unless the prototype pressure is

**TABLE 6.1  Exponents for Pressure Scale Effects Adjustments on Incipient and Critical Cavitation for Valves**

| Valve, type<br>(1) | Exponent for Eq. 6.6<br>$X$<br>(2) | | Range of Pressure for<br>Experimental Data<br>kPa<br>(3) |
|---|---|---|---|
| 4 in. butterfly | 0.28 | | 120–660 |
| 6 in. butterfly | 0.28 | | 140–1310 |
| 12 in. butterfly | 0.28 | | 120–1200 |
| 12 in. butterfly | 0.28 | | 100–930 |
| 12 in. butterfly | 0.24 | | 80–930 |
| 20 in. butterfly | 0.30 | | 70–550 |
| 24 in. butterfly | 0.24 | | 150–740 |
| Average | | 0.28 | |
| 2 in. ball | 0.30 | | 120–1700 |
| 8 in. ball | 0.28 | | 50–1030 |
| 12 in. ball | 0.24 | | 100–620 |
| Average | | 0.27 | |
| 24 in. cone (skirted)[a] | 0.22 | | 450–1520 |
| 16 in. globe[a] | 0.14 | | 450–1340 |
| 8 in. Pelton needle | 0.14 | | 450–1030 |

[a]Valves tested by Metropolitan Water District of Southern California, all other valves were tested by the author or his research associates.

much larger than the experimental value. In the absence of experimental pressure scale effects data on a specific valve, scale effects caused by difference in pressure between model and prototype can be approximated with average values of $X$ obtained from Table 6.1.

Values of $X$ for the globe and needle valves listed in Table 6.1 are somewhat lower than for the butterfly or ball valves. This indicates less pressure scale effect since when $X = 0$, pressure scale effects are absent. The reason for the decreased scale effect is associated with details of the valve geometry near the seat. (It will be discussed in Chapter 7 that for orifices, no pressure scale effects exists i.e. $X = 0$.) The jet from needle and globe valves approaches that for an orifice. In both cases, the jets discharge into a sudden enlargement and are completely surrounded by liquid. This provides optimum conditions for suppressing cavitation and minimizes pressure scale effects.

Most of the scale effects work for *incipient damage* has been done with orifice plates (59, 60) because of the ease in maintaining geometric similarity. As was found for incipient and critical cavitation, the pressure scale effects for $\sigma$ at incipient damage follow the equation

$$\sigma_{id} = \mathrm{PSE}\,\sigma_{ido} \qquad (6.8)$$

where PSE is calculated from Eq. 6.7a or b. Values of the exponent $X$ for orifices, butterfly, and globe valves are found in Table 6.2.

Considerable research has shown that there are no significant size or pressure scale effects for choking cavitation. This would be expected because the mean pressure at the vena contracta can be predicted with the energy equation if the contraction coefficient is known. It is therefore possible to calculate $\sigma_{ch}$ for some devices. The factors causing the scale effects for the other limits do not apply at choking. Therefore, the choking index $\sigma_{ch}$ obtained on a model at any pressure can be used without adjustment for predicting prototype performance.

TABLE 6.2   Exponents for Pressure Scale Effects Adjustments on Incipient Damage for Valves and Orifices

| Device (1) | Exponent for Eq. 6.7 $X$ (2) | Range of Pressures for Experimental Data kPa (3) |
|---|---|---|
| Orifice | 0.19 | 340–1300 |
| Butterfly valve | 0.18 | 270–1000 |
| Globe valve | 0.30 | 240–690 |

**Size Scale Effects**

The primary research work to evaluate size scale effects was done using orifices (18, 82). This work was checked against size scale effects data obtained on several types of valves. The resulting equation recommended for making size scale effects adjustments (SSE) on $\sigma_i$ or $\sigma_c$ is (Ref. 46a)

$$SSE = \left(\frac{D}{d}\right)^Y \tag{6.9}$$

in which $D$ is the valve diameter and $d$ is the reference valve diameter. The exponent $Y$ is evaluated by

$$Y = 0.3 K_l^{-0.25} \tag{6.10}$$

in which $K_l$ is defined by Eq. 4.1. The scale effect adjustment is made by

$$\sigma_c = SSE\sigma_{co} \quad \text{and} \quad \sigma_i = SSE\sigma_{io} \tag{6.11}$$

where $\sigma_{co}$ and $\sigma_{io}$ are reference cavitation data for a valve of size $d$. These equations are valid for orifices and valves.

Equations 6.9–6.11 has been checked for valves up to 36-in. diameter. For larger valves, there is no experimental information available to assist in making size scale effects adjustments. It is this researcher's opinion that the size scale effects on valves larger than 3-ft (1-m) diameter would be smaller than predicted by Eq. 6.11. This is based on the fact that for most hydraulic modeling situations involving size or Reynolds number scale effects, there is a size or upper limit of Reynolds number beyond which the size scale effects are negligible. For example, recall the variation of the friction factor $f$ with Reynolds number. At large Re, $f$ becomes independent of Re. This would also seem to apply to cavitation. In the absence of any further information, however, reasonable engineering judgment must be used in determining how much to reduce the size scale effects for large valves.

Equations for making pressure scale effects adjustments and size scale effects were developed independently. Their effects can be combined.

$$\sigma_c = PSE \cdot SSE\sigma_{co}$$

or $\qquad\qquad\qquad\qquad\qquad\qquad\qquad\qquad$ (6.12)

$$\sigma_i = PSE \cdot SSE\sigma_{io}$$

Experimental data on size scale effects for incipient damage are available only for orifices. For size scale effects on incipient damage based on the pitting rate, the data (46, 59, 60) show that orifices in three different pipe sizes had the same value $\sigma_{id}$, that is no size scale effect. However, the research (59) clearly showed that the size of the pits significantly increased with pipe diameter. Therefore, once material begins to be removed by cavitation, the larger systems will produce more weight loss than would be predicted from model tests.

**Example Problems**

Several simple examples are presented to demonstrate use of the scale effects equations and experimental data. These examples are limited to analyses of given situations. More complex examples which demonstrate how to apply the information to design systems are given in Section 6.4.

*Example 6.1.* Using the following laboratory data obtained on a 101-mm butterfly valve ($C_d = 0.083$, $K_l = 144$, $P_{do} = 102$ kPa, $\sigma_{co} = 1.14$), predict the critical cavitation conditions for a similar 610-mm butterfly valve at $P_d = 698$ kPa. $P_{vg} = P_{vgo} = -77$ kPa. Also find the percent error if scale effects are ignored.

From Eq. 6.9, the size effects are SSE $= 1.16$ and from Eq. 6.7a, the pressure scale effects are (use average $X$ from Table 6.1):

$$\text{PSE} = [(698 + 77)/(102 + 77)]^{0.28} = 1.51$$

From Eq. 6.12,

$$\sigma_c = 1.14 \cdot 1.51 \cdot 1.17 = 2.01$$

If scale effects were ignored, $\sigma_c$ would be 1.14 for the large valve (the same as for the test valve). The percent error in $\sigma_c$ would be 44%.

*Example 6.2.* A 6-in. butterfly valve is to be installed and operated under the following conditions: $P_u = 104$ psi, $P_d = 80.8$ psi, $P_{va} = 1.16$ psi, $Q = 1.29$ cfs, $P_b = 12.2$ psi. Determine if the valve will cavitate.

The procedures will involve comparing the value of $\sigma_{\text{system}}$ with the value of $\sigma_c$ or $\sigma_i$ from Fig. 6.4 or Eq. 6.3b. To get $\sigma_c$ and $\sigma_i$ requires knowing $C_d$ and adjusting for pressure scale effects.

$$P_u - P_d = 104 - 80.8 = 23.2 \text{ psi or } 23.2 \cdot 2.308 = 53.5 \text{ ft}$$

$$V = \frac{Q}{A} = 1.29/0.196 = 6.58 \text{ fps}$$

$$C_d = \frac{6.58}{(64.4 \cdot 53.5 + 6.582^2)}^{0.5} = 0.111$$

$$\sigma_{\text{system}} = \frac{(80.8 + 12.2 - 1.16)}{23.2} = 3.95$$

From Fig. 6.4 at $C_d = 0.111$, or from Eq. 6.3a and b, $\sigma_c = 1.67$ and $\sigma_i = 2.73$, for $P_{uo} = 70$ psi and $P_{vgo} = -12$ psi.

Adjust for pressure scale effects with $P_{vg} = 1.16 - 12.2 = -11$ psi and $X = 0.28$ (Table 6.1).

$$\text{PSE} = [(104 + 11)/(70 + 12)]^{0.28} = 1.10$$

$$\sigma_i = 1.10 \cdot 2.78 = 3.06$$

Since $\sigma_{\text{system}} > \sigma_i$, there will be no cavitation.

***Example 6.3.*** A 2-m butterfly valve is to be installed in a pipeline where the gauge pressure upstream from the valve will be $P_u = 800$ kPa. The valve will operate at $C_d = 0.5$. Determine $\sigma_c$ and the allowable velocity ($V_c$) through the 2-m pipe so the valve operates at critical cavitation ($P_{vg} = -89.6$ kPa). Also find the percent error in $V_c$ if scale effects are ignored.

The scale effects in this example will be primarily due to variations in size. Assume the prototype valve closely resembles the 152-mm (6-in.) butterfly valve in Fig. 6.4. Therefore, both the discharge coefficient and cavitation data for the small valve can be used to predict performance of the large valve. The experimental data for the small butterfly valve (from Fig. 6.4 or Eq. 6.3b and Table 6.1) are

$$X = 0.28 \qquad\qquad C_d = 0.5$$

$$D = 2\text{m} \qquad\qquad P_{uo} = 482 \text{ kPa (70 psi)}$$

$$d = 0.152\text{m} \qquad\qquad P_{vgo} = -82.7 \text{ kPa}$$

$$\sigma_{co} = 5.1 \text{ (Fig 6.4)} \qquad K_l = 3.0$$

Scale $\sigma_{co}$ for size and pressure scale effects;

$$Y = 0.3(3.0)^{-0.25} = 0.228$$

$$\text{SSE}\left(\frac{2}{0.152}\right)^{0.228} = 1.80$$

$$\text{PSE} = [(800 + 89.6)/(482 + 82.7)]^{0.28} = 1.14$$

$$\sigma_c = \text{PSE}\sigma_{co} \cdot \text{SSE} = 5.1 \cdot 1.14/1.8 = 10.5$$

The allowable velocity can be calculated by determining the pressure drop corresponding to $\sigma_c = 10.5$ and using $C_d$ or any of the other flow coefficients:

$$P_d = (\sigma_c P_u + P_{vg})/(1 + \sigma_c) = 732 \text{ kPa}$$

$$\Delta P = 800 - 732 = 68 \text{ kPa}$$

$$V_c = (2\Delta P/\rho\ K_l)^{0.5} = 6.73 \text{ m/s}$$

If scale effects are ignored, $\sigma_c = 5.1$, $P_d = 654$ kPa, $\Delta P = 800 - 654 = 146$ kPa and $V_c = 9.86$ m/s. The percent error in $V_c$ would be $100(9.86 - 6.73)/6.73 = 46\%$

***Example 6.4*** (***Choking Cavitation***). A 4-in ball valve is to be used for pressure relief. The line pressure is 72.5 psi and the valve discharges into a tank which provides a back pressure of 7.25 psi; assume that $C_d = 0.808$. Determine if the valve is choking and compute the discharge for the valve operating at 90° open ($P_{vg} = -13$ psi).

Assume that the valve is similar to the 2-in. ball valve shown in Fig. 6.8. To determine the degree of cavitation, use one of the five methods listed in Section 6.2. For this example, $\sigma_{system}$ will be calculated and compared with the various cavitation limits. With such a large pressure drop across the valve, it will likely be choking. Since there are no size or pressure scale effects at choking, no scale effect adjustments are necessary.

$$\sigma_{system} = (7.25 + 13)/(72.5 - 7.25) = 0.31$$

From Fig. 6.8 for the 2-in. valve at $C_d = 0.808$ (90° open), $\sigma_{ch} = 4.1$. Since $\sigma_{ch}$ is much greater than $\sigma_{system}$, the valve will be operating beyond choking, in the supercavitation range.

The coefficient of discharge for the valve is 0.808 at 90° open. This corresponds to nonchoking conditions and cannot be used with the system to calculate discharge for a valve at choking or operating in supercavitation. The correct procedure would be as follows. From Eq. 6.1,

$$\sigma_{ch} = 4.1 = (72.5 - \Delta P_{ch} + 13)/\Delta P_{ch} = 16.8 \text{ psi} = 38.8 \text{ ft}$$

For any $\Delta P > 16.8$ psi, the valve will be choked. Note that at $Q_{ch}$ the equations for $C_d$, etc., are valid and can be used to calculate $V_{ch}$.

$$Q_{ch} = AV_{ch}, \quad V_{ch} = (64.4 \cdot 38.8/K_l)^{0.5} K_l = 1/0.808^2 - 1 = 0.532$$
$$Q_{ch} = 0.087 \cdot 68.6 = 5.96 \text{ cfs}$$

Velocities of this magnitude would not generally be allowed in the pipe because of cavitation at elbows or other minor losses. If the 5.93 cfs were needed, it would be better to use a larger pipe and reduce at the valve. It would also be necessary to be sure that the valve could handle the torque caused by that high velocity.

If the system $\Delta P$ of $72.5 - 7.25 = 65.2$ psi were used with $K_l$ to calculate $V_{ch}$ and $Q_{ch}$ the result would be $V_{ch} = 135$ fps and $Q_{ch} = 11.7$ cfs. This is 97% higher than the actual flow rate.

## 6.4   CONTROLLING CAVITATION

In the previous sections, information was provided on various type of valves that can be used to identify the conditions under which they will operate at various levels of cavitation. Examples were given to demonstrate how to apply the data to determine the intensity of cavitation that a valve will be subjected to when operating at a given set of system conditions. If such an analysis indicates that the valve, orifice, or other device will be operating at a cavitation level greater than can be tolerated, the design must be altered. This section provides information on various techniques which can be used to provide a final design which can operate at the desired level of cavitation; the various techniques will be described and examples provided.

### Type of Valve

By carefully examining the cavitation data provided in Figs. 6.4–6.7, for various types of valves, it can be seen that there is a measurable difference in their cavitation performance. The normal procedure in selecting a valve for a particular installation is to choose the most economical valve that will satisfy the system requirements. Comparing the data in Figs. 6.4–6.7 demonstrates the differences in the cavitation performance of different valves. For example, compare the 24 in. solid and skeleton cone valves in Fig. 6.6. The $\sigma_c$ for the skeleton cone valve is as much as double that of the solid cone valve, showing that it cavitates much easier. A similar comparison can be made between the two globe valves in Fig. 6.7. By selecting the proper valve it may be possible to avoid cavitation in certain situations.

If using conventional valves such as butterfly, cone, ball, or globe valves, still produces excessive cavitation, it may be advisable to look at new valve styles developed in recent years. Some valves on the market can produce high pressure drops under controlled cavitation conditions. These valves operate generally on one of two principles. One class of valves uses the principle of flow through an orifice discharging into an infinite enlargement. The orifice discharging into a downstream enlargement has been found to be a good arrangement for suppressing cavitation. The reason is that the cavitation occurs in the fluid, away from solid boundaries. There are several valves on the market which use this principle (80). One is the sleeve valve. Another is the submerged Howell–Bunger valve (Fig. 4.1).

The other principle that is utilized in valves for obtaining high pressure drops under controlled cavitation conditions is to dissipate the energy by placing a number of constrictions in series. This type of valve can be best visualized by comparing it to a number of orifice plates in series. By properly dividing the pressure drop between the orifice plates, one can increase the allowable pressure drop almost without limit and still control the cavitation.

**Valve in Parallel or Larger Valves**

If a single valve cannot produce the required pressure drop without cavitation, one solution may be to use a larger valve or place two valves in parallel. To explain how this helps, first consider the cavitation characteristics of valves. Referring to Figs. 6.4–6.7, it is observed that for these cavitation limits, the magnitude of $\sigma$ increases significantly as the valve opening increases. Since the magnitude of $\sigma$ is inversely related to the allowable pressure drop across the valve, a valve operating at a smaller opening can produce a larger pressure drop at the same level of cavitation. Consequently, if a larger valve is used, or two valves are placed in parallel, the discharge through each is reduced and each valve operates at a smaller valve opening and a smaller $C_d$ to produce the same pressure drop. The decision whether to use this technique, or others that will be described, is primarily based on economy, ease of operation, and whether the valve is forced to operate at too small an opening. One advantage of the two-valve scheme is that this technique provides more reliability because there are two lines in parallel, each of which can supply part or possibly the entire required flow under emergency conditions.

One of the problems with using a larger valve or valves in parallel, besides increased cost, is the limitation of not operating valves at extremely small valve openings. It is generally accepted that a valve should not be used for throttling much below 10–15% open. Difficulties arise due to high pressure drops causing damage to the seats and the poor flow control. These problems become even more difficult if a wide range of discharge is required. One solution to the problems is to place valves of varying size in parallel. Each line can be designed to operate cavitation-free over a portion of the required discharge range.

*Example 6.5.* Flow between two reservoirs is provided with a 508-mm diameter pipe. The supply reservoir is at an elevation of 1300 m. The receiving reservoir is at an elevation of 1270 m. The pipe is 500 m long, $f = 0.013$, and a valve structure is to be built near the receiving reservoir at an elevation of 1240 m. Water temperature varies from 7 to 20°C. Flow rates are required between 0.3 and 1.0 m$^3$/s. (Only calculations for $Q_{max}$ will be included.) Design a valve installation that will operate at or below critical cavitation.

First try to select a single valve that will operate below critical cavitation:

$$P_d = (1270 - 1240) = 30 \text{ m}, \quad V = Q/A = 4.93 \text{ m/s}$$

$$P_u = (1300 - 1240) - fLV^2/2Dg = 44.1 \text{ m}$$

$$\Delta P = 44.1 - 30 = 14.1 \text{ m}$$

$$C_d = 4.93/(2 \cdot 9.81 \cdot 14.1 + 4.93^2)^{0.5} = 0.284$$

$$\sigma = (P_d + P_{atm} - P_{va})/(P_u - P_d)$$

From Table 1.1, $P_{va} = 0.22$ m of water $= 2.2$ kPa at 20° C. From Eq. 1.11, at elevation $= 1240$ m (4000 ft), $P_{atm} = 12.7$ psi $= 8.93$ m of water, $P_{vg} = 0.22 - 8.93 = -8.71$ m.

$$\sigma_{system} = (30 + 8.71)/14.1 = 2.74$$

Using the data in Figs. 6.4–6.7, try to find a valve that can meet these requirements. Remember that globe valves are not made larger than 406 mm.

(a) First try butterfly valves: use the data from Fig. 6.4. At $C_d = 0.284$, Eq. 6.3b gives $\sigma_c = 3.06$, compared with $\sigma = 2.79$ required by the system. Since scale effects will increase this value, the valve will cavitate.

(b) Next try to solve the problem by using the 610-mm solid cone valve in Fig. 6.5. At $C_d = 0.284$, Eq. 6.3b gives $\sigma_c = 2.02$, at $P_u = 52.7$ m (75 psi). Adjusting for PSE and SSE will reduce $\sigma_c$ so the solid cone valve will be satisfactory since $\sigma_c < \sigma_{system}$.

As an alternative solution, consider using butterfly valves in parallel to reduce the velocity in each line, and therefore, allow them to operate at lower openings and have lower $\sigma_c$.

Since the velocities must be decreased to make the concept work, the combined area of the valves must be greater than 0.203 m$^2$ (area of one 508-mm valve). Try two 508-mm butterfly valves, $A = 2 \times 0.203 = 0.406$ $m^2$ at $Q_{max}$ (with the flow equally divided):

$V = 2.46$ m/s (in each 508-mm line)

$P_d = 30$ m, $P_u = 44.1$ m (ignoring losses where pipe divides)

$C_d = 0.146$

From Eq. 6.3b at $C_d = 0.146$, $K_l = 45.9$, $\sigma_{co} = 1.98$ for the 153-m butterfly valve at $P_u = 49$ m, and $P_{vgo} = -8.43$ m. Using Eqs. 6.9–6.11 SSE $= 1.15$, so $\sigma_c = 2.28$

Since $\sigma_{co} < \sigma_{system}$ at maximum flow, there will be no cavitation. $Q_{min}$ and several conditions in-between would still need to be checked to be sure there are no problems at other flow conditions.

**Valves in Series**

Another technique for suppressing cavitation is to place valves in series, thus reducing the pressure drop across each valve and allowing them to operate at a larger $\sigma$. This technique is probably more efficient than using valves in parallel, but it significantly increases the complexity of the controls. To take maximum advantage of the two valves in series, the upstream valve should always operate at smaller valve openings since it can take a larger pressure drop. Because the two valve positions depend on each other and on the flow rates, the control system must be able to keep the proper valve opening for each valve in order to control cavitation.

Proper spacing of valves placed in series is important. The spacing between valves depends on the type of valve. For butterfly valves, 5–8 pipe diameters are required between valves to prevent flutter of the leaf of the downstream valve and obtain the normal pressure drop across each valve. For globe, cone, and other types of valves, it is possible to bolt them flange to flange and have satisfactory operation based on head loss. One must be concerned with the effect of the upstream valve causing cavitation damage on the downstream valve and possible instability. If the valves are designed to operate at less than incipient damage, this does not become a design criterion. However, if the valves will be operating in a cavitation region, bolting valves flange to flange would not be advisable.

*Example 6.6.* To provide minimum stream flow from a reservoir, an outlet pipe with a valve structure is required. The normal reservoir head provides a shutoff pressure at the valve location of 71 psi (164 ft). The discharge pipe is short and empties into the atmosphere; therefore assume $P_d = 0$. The minor and friction losses in the upstream pipe can be ignored. Assume the maximum flow required is 11.7 cfs. Design the structure using globe valves in series and have them operate below incipient damage. Assume $P_b = 13.0$ psi, $T_{max} = 40°F$.

First check to see if one valve might be sufficient. From Table 1.1, $P_{va} = 0.28$ ft $= 0.12$ psi.

$$\sigma_{system} = \frac{(0 + 13 - 0.12)}{(71 - 0)} = 0.181$$

Checking the data in Fig. 6.7 for globe valves shows that $\sigma_{system} < \sigma_{id}$ for any valve opening. Therefore, it will require two or more valves in series.

Select a valve and pipe diameter so that when the valves are wide open the flow will be at least 11.7 cfs. Assume 3 valves and apply the energy equation from the reservoir to the outlet where the pressure is zero (ignoring friction and minor losses).

$$164 = 3K_l \frac{V^2}{2g}$$

From Fig. 6.7 (lower set of data), $C_d = 0.378$ at full open and $K_l = (1/C_d^2) - 1 = 6$. Substituting $K_l = 6$ into the energy equation gives $V = 24.2$ fps. This corresponds to a valve diameter of 9.4 in.

Since only available pipe and valve sizes can be used, either a 10- or 12-in. valve can be used; however, select 12-in. valves because a larger valve will have better cavitation characterictics. The three 12-in. globe valves will have to produce a pressure drop of 71 psi at a flow of 11.7 cfs.

In evaluating the cavitation performance of the globe valves and deciding how the required pressure drop will be divided between them, it is important to realize that the upstream valves can take more pressure drop. This is because the downstream valves create a higher back pressure and allow the upstream valves to produce a higher pressure drop and therefore operate at a lower valve opening. Valves have considerably better cavitation performance at smaller openings, so this even further increases the allowable pressure drop. The first set of calculaions will be done by hand to demonstrate the process. Subsequent calculations will be done with a spreadsheet.

Start the cavitation analysis at the downstream valve by assuming a pressure upstream of the downstream valve. For the first set of calculations:

*Valve #1*

$$P_u = 25 \text{ psi (assumed)}$$

$$P_d = 0 \text{ (given)}$$

$$P_{vgo} = -12 \text{ psi (Fig 6.7)}$$

$$P_{uo} = 60 \text{ psi (Fig 6.7)}$$

$$V = 14.9 \text{ fps}$$

$$C_d = \frac{14.9}{(64.4 \cdot 2.308 \cdot 25 + 14.92^2)} = 0.237$$

$$\sigma_{\text{system}} = \frac{(0 + 13 - 0.12)}{25} = 0.52$$

$$\sigma_{ido} = 0.99 \text{ (Eq. 6.4f or Fig. 6.7)}$$

$$P_{do} = \frac{(\sigma_{ido} P_{uo} + P_{vgo})}{(1 + \sigma_{ido})} = 23.8 \text{ psi}$$

Make the PSE adjustment using Eqs. 6.6 and 6.7. No other scale effects adjustments are needed since there are no SSE adjustments for $\sigma_{id}$.

$$X = 0.3 \text{ (from Table 6.2.)}$$

$$\text{PSE} = [(0 + 13 - 0.12)/(23.8 + 12)]^{0.3} = 0.74$$

$$\sigma_{id} = 0.74 \cdot 0.99 = 0.73$$

Since $\sigma_{id} > \sigma_{system}$, the valve will be operating in damaging cavitation. The assumed pressure drop was too large. The calculations need to be repeated until $\sigma_{id} < \sigma_{system}$. A spreadsheet is ideal for this. Table 6.3 lists the output of a spreadsheet application that shows the solution to this problem. Note that the first line of data (row 3) lists the data just calculated. The data entered in each cell of the spreadsheet are tabulated at the bottom. The spreadsheet can automatically iterate to converge on the solution, but it is more instructive to show the separate calculations.

The second guess for $P_u$ of 10 psi is too small, since $\sigma_{system}$ is considerably less than $\sigma_{id}$. The third guess of $P_u = 12$ psi is satisfactory.

Calculations for the second valve start with $P_d = 12$ psi ($P_u$ for valve #1). Again, three calculations are listed. For the third valve, $\sigma_{system}$ is considerably larger than $\sigma_{id}$. This confirms that three 12-in. globe valves will be adequate. The final design for $Q = 11.7$ cfs is found in rows 5, 9, and 11.

Note that pressure drop across the third valve is not arbitrary. The upstream pressure must be 71 psi and $P_d$ must equal $P_u$ for valve #2. Valve #3 will operate well out of cavitation.

Discussion:
1. It might be advisable to reduce the pressure drop across valves #1 and #2 and increase it across valve #3 so they all operate farther away from damage.
2. At the design discharge of 11.7 cfs, the three valves in series can operate at or below incipient damage
3. To complete the analysis, the cavitation level must be checked at several flow rates between maximum and minimum flow. The procedure would be identical to that used at maximum flow.
4. One of the complexities of this kind of a solution (valves in series) is that each valve operates at a different opening. This requires a more complex operating system than a single valve or valves in parallel.
5. Since globe valves are used in this case, automatic hydraulic control might be the most satisfactory. With simple controls, globe valves can automatically maintain a constant upstream or downstream pressure or maintain a certain flow rate.

**TABLE 6.3 Spreadsheet for Example 6.6**

| | A | B | C | D | E | F | G | H | I | J | K | L | M | N | O | P | Q |
|---|---|---|---|---|---|---|---|---|---|---|---|---|---|---|---|---|---|
| 1 | DEVICE | Flow | ID | $P_u$ | $P_d$ | $V$ | $P_{vg}$ | $\sigma$ | $P_{vgo}$ | $C_d$ | $\sigma$ | $P_{uo}$ | $P_{do}$ | SSE | $X$ | PSE | $\sigma$ |
| 2 | | cfs | in. | psi | psi | fps | psi | sys. | psi | | ido | psi | psi | | | | id |
| 3 | #1, 12″ globe | 11.7 | 12.0 | 25.0 | 0.0 | 14.9 | −12.9 | **0.52** | −12.0 | 0.237 | 0.99 | 60.0 | 23.8 | 1.00 | 0.30 | 0.7 | **0.73** |
| 4 | #1, 12″ globe | 11.7 | 12.0 | 10.0 | 0.0 | 14.9 | −12.9 | **1.29** | −12.0 | 0.360 | 1.56 | 60.0 | 31.9 | 1.00 | 0.30 | 0.71 | **1.11** |
| 5 | #1, 12″ globe | 11.7 | 12.0 | 12.0 | 0.0 | 14.9 | −12.9 | **1.08** | −12.0 | 0.333 | 1.48 | 60.0 | 31.0 | 1.00 | 0.30 | 0.73 | **1.08** |
| 6 | | | | | | | | | | | | | | | | | |
| 7 | #2, 12″ globe | 11.7 | 12.0 | 25.0 | 12.0 | 14.9 | −12.9 | **1.92** | −12.0 | 0.321 | 1.44 | 60.0 | 30.4 | 1.00 | 0.30 | 0.82 | **1.18** |
| 8 | #2, 12″ globe | 11.7 | 12.0 | 30.0 | 12.0 | 14.9 | −12.9 | **1.38** | −12.0 | 0.277 | 1.22 | 60.0 | 27.6 | 1.00 | 0.30 | 0.85 | **1.04** |
| 9 | #2, 12″ globe | 11.7 | 12.0 | 43.0 | 12.0 | 14.9 | −12.9 | **0.80** | −12.0 | 0.214 | 0.85 | 60.0 | 21.1 | 1.00 | 0.30 | 0.93 | **0.79** |
| 10 | | | | | | | | | | | | | | | | | |
| 11 | #3, 12″ globe | 11.7 | 12.0 | 71.0 | 43.0 | 14.9 | −12.9 | **2.00** | −12.0 | 0.225 | 0.91 | 60.0 | 22.4 | 1.00 | 0.30 | 1.05 | **0.96** |
| 12 | | | | | | | | | | | | | | | | | |
| 13 | Column | Source of data | | | | | | | | | | | | | | | |
| 14 | A | Input | | | | | | | | | | | | | | | |
| 15 | B | Given data | | | | | | | | | | | | | | | |

| 16 | C | Calculated during preliminary calculations |
| 17 | D | Assumed |
| 18 | E | Initially given as 0, for upstream valves it equals $P_u$ of downstream valve |
| 19 | F | $Q/$area |
| 20 | G | Calculated during preliminary calculations |
| 21 | H | Calculated system $\sigma = (P_d - P_{vg})/(P_u - P_d)$ |
| 22 | I | Found from reference data (Fig. 6.7 for this example) |
| 23 | J | $\dfrac{V}{\sqrt{2g(P_u - P_d)2.308 + V^2}}$ |
| 24 | K | Reference cavitation data (Eq. 6.4f used for this example) |
| 25 | L | Found from reference data (Fig. 6.7 for this example) |
| 26 | M | $[\sigma(ido) \cdot P_{uo} + P_{vgo}]/(1 + \sigma(ido)]$ |
| 27 | N | Normally use Eq. 6.9, but since there are no SSE for incipient damage, enter 1.0 |
| 28 | O | From Table 6.1 or 6.2 |
| 29 | P | Calculated with Eq. 6.7a for PSE |
| 30 | Q | Calculated with Eq. 6.12 for PSE and SSE |

### Cavitation Resistant Materials

Another method of suppressing cavitation damage is to cover parts of the valve and downstream piping with cavitation resistant materials. With this solution, one tolerates the cavitation and tries to prevent erosion. The advantage of this method is that the simplicity of a single valve controlling the flow is maintained. The disadvantage is the cost. A cost analysis should be made to determine which of the various methods is most satisfactory.

Table 6.4 (29) lists the relative resistance of numerous types of materials to cavitation erosion damage. These data were obtained from tests using a magnetostriction device, whereby cavitation is created by a very high-frequency vibration of a small button of material. The weight loss in milligrams over a 2-h period is listed for the materials. It can be seen that there is a wide variation in the resistance of the various types of materials. One caution that should be mentioned regarding application of the data in Table 6.4 is that different types of cavitation may result in different erosion rates. With the magnetostriction device, the cavitation events are very small and occur at an extremely high rate since the device is vibrating at approximately 20,000 cycles per second. If the same material is subjected to cavitation occurring in a flowing system such as valves or pumps, the cavitation events are much larger and occur at a lower frequency, where the

**TABLE 6.4  Cavitation Erosion Resistance of Metals**[a]

| Alloy | Magnetostriction Weight Loss After 2 h (mg) |
|---|---|
| Rolled stellite[b] | 0.6 |
| Welded aluminium bronze | 3.2 |
| Cast aluminum bronze | 5.8 |
| Welded stainless steel (2 layers, 17 Cr-7 Ni) | 6.0 |
| Hot rolled stainless steel (26 Cr-13 Ni) | 8.0 |
| Tempered rolled stainless steel (12 Cr) | 9.0 |
| Cast stainless steel (18 Cr-8 Ni) | 13.0 |
| Cast stainless steel (12 Cr) | 20.0 |
| Cast manganese bronze | 80.0 |
| Welded mild steel | 97.0 |
| Plate steel | 98.0 |
| Cast steel | 105.0 |
| Aluminium | 124.0 |
| Brass | 156.0 |
| Cast iron | 224.0 |

[a] Reference 29.

[b] Despite the high resistance of this material to cavitation damage, it is not suitable for ordinary use because of its comparatively high cost and the difficulty encountered in machining and grinding.

(Courtesy Interscience Publishers, Div. of John Wiley & Sons.)

**TABLE 6.5    Incipient Cavitation Damage Data for Different Metals**[a]

| Material | $P_u$ (psi) | $\sigma_{id}$ | $V_{id}$ (fps) |
|---|---|---|---|
| Aluminium | 80 | 0.804 | 15.8 |
| Brass | 80 | 0.512 | 16.9 |
| Carbon steel | 81 | 0.397 | 17.8 |
| 304 stainless steel | 80 | 0.404 | 17.6 |

[a]Reference 46. (Courtesy W. J. Rahmeyer.)

different materials may behave in a substantially different manner. Limited laboratory work and some field observations indicate that the relative resistance of materials to cavitation in real situations is less than indicated by the numbers in Table 6.4.

Limited testing has been done on the erosion resistance of different materials and coatings to cavitation in flowing systems. Reference 46 reports on the relative resistance of aluminum, brass, 304 stainless steel, and carbon steel subjected to cavitation caused by an orifice. Cavitation was detected by counting the pitting rate at different levels of cavitation. The experimental data are listed in Table 6.5. The data show that basing incipient damage conditions on tests with soft aluminum is conservative and also demonstrates that there is less variation in the damage resistance of materials in a flowing system. This conclusion is based on pitting, not weight removal, so it may not be a direct comparison with the data in Table 6.4.

### Aeration

Injecting air to suppress cavitation is a technique that has been used for many years with varying degrees of success. If an adequate amount of air is injected into the proper region, noise, vibrations, and erosion damage can be significantly reduced. If the air is introduced so that it gets dispersed throughout the region where cavitation originates, it suppresses the formation of vaporous cavitation and cushions the collapse. When cavitation nuclei are made up of a large percentage of gas, vaporization of the fluid is greatly reduced due to expansion of the gas. This type of cavitation is called gaseous cavitation and usually does not cause serious damage.

In the zone where cavitation collapse occurs, a large amount of air provides a cushioning effect, reducing noise, vibration, and the possibility of erosion damage. This is due to the reduced wave speed of the liquid. With a lower wave speed, the magnitude of the shock wave created when the cavity collapses is much less. This reduces the damage potential.

The concept of using air to suppress cavitation can be applied both to open-channel and closed-conduit flow. There are a few fundamental differences between injecting air into a pipe to suppress cavitation at a valve, orifice, pump, or turbine and using an air slot to suppress cavitation on a spillway. One difference is that in a pipe, the cavitation is usually local, being generated at a valve, pump, turbine, or other minor loss. As a result, only a small portion of the system must be protected from cavitation erosion damage. However, with an open-channel application such as a spillway, large areas of the surface must often be protected.

For air slots on spillways or at gate slots, it is usually desirable to maximize the amount of air drawn in by the air slot. This is desirable in order to keep the concentration of air at the boundary high for a greater distance. Since buoyant forces and pressure gradients cause the air to rise to the surface, an air slot can only protect the boundary for a finite length. It may be necessary to place air slots in series.

For pipe flow, one must be concerned about the effect of the free air on the accuracy of flow meters, reduction of flow due to blockage, transients caused by large slugs of air moving through the pipe or through a control valve, plugging of filters at water treatment plants, and various other similar problems. It is therefore necessary for closed conduit flow to minimize the amount of air required to suppress cavitation.

When using air to suppress cavitation at a valve, pump, turbine, etc., it is difficult to identify all of the zones of cavitation and to supply air to those areas. Often, cavitation in a closed-conduit system will occur when the mean pressure in the cavitation region is above atmospheric. This necessitates the use of compressed air. Such a solution is possible, but it is more complex and costly and less reliable. The amount of air required to suppress cavitation is generally between 2 and 6% by volume of the water flow rate. This figure is based on a study with orifice plates (11), a cone valve (40) and a butterfly valve (15). The percentage for other valves may be different.

One of the most common mistakes when attempting to reduce cavitation by injection air is placing the air inlet port in the wrong location so the air does not get to the cavitation zone. The author is aware of an installation where air was supplied to a 36-in. butterfly valve to suppress cavitation. The valve was installed with the shaft vertical and two 3-in. air ports were located on the sides of the pipe so they were in a plane perpendicular to the shaft. The air ports sucked in large quantities of air but did not reduce the cavitation. The problem was that the air was drawn in by the high-velocity jets springing from each side of the leaf and washed downstream. It was not able to penetrate through the jets and get into the separation zone in the shadow of the disk. When the air ports were placed on the top and bottom of the pipe close to the shaft, the cavitation was suppressed.

A similar problem was observed in the laboratory with a skeleton ball valve. The air port was installed on the side of the pipe close to the discharge flange and on the opposite side from the outlet port. The valve sucked in air, but the cavitation was not reduced. The problem was that there was enough water flowing around the ball that it washed the air downstream and prevented it from getting to the proper region.

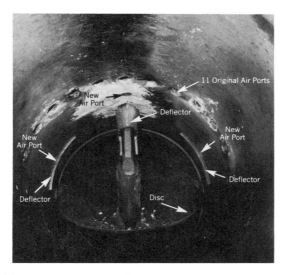

**Fig. 6.10**  Looking upstream at a 48-in. pivot butterfly valve with aeration ports. (Photo courtesy of Freese and Nichols, Fort Worth, Texas.)

For some installations, it may be difficult to place the air ports in the optimum location. Figure 6.10 shows a 48-in. butterfly valve installation that was plagued with serious cavitation problems. The original design included 11 air inlet ports spaced around the top half of the pipe and located about 4 ft downstream. These sucked in a lot of air but did not reduce the cavitation because the air did not get into the cavitation zone.

Later, three additional air ports were installed. One 1-in. diameter hole was located on the top of the pipe just downstream from the deflector. The other two (2-in. dia.) were placed on the sides closer to the valve. The construction of the valve prevented the air ports from being directly in line with the valve shaft. To be sure that the air would make its way into the separation region, 1-in. high deflectors were welded to the pipe, as shown in Fig. 6.10. The operating shaft deflector was also modified to improve aeration. With these alterations, the cavitation was almost totally eliminated.

Injecting air may not entirely eliminate the cavitation damage. Table 6.6 lists cavitation aeration data on $\sigma_{id}$ for a skirted cone valve and a butterfly valve. The percent reduction in $\sigma_{id}$ when air was introduced varied between about 10 and 60%. The reduction is significant enough to consider injecting air.

*Aeration scale effects.* Our present understanding of cavitation does not furnish the tools necessary to predict the amount of air required to suppress cavitation without experimental data. Because of modeling problems, obtaining good experimental data on aeration and cavitation is difficult. When

**TABLE 6.6  Influence of injecting air on $\sigma_{id}$.**

| V.O. deg | $C_d$ | $\sigma_{id}$ | $\sigma_{id}$ (air) | $\% \Delta\sigma$ | %Air |
|---|---|---|---|---|---|
| 6″ butterfly valve (ref. 14a) | | | | | |
| 20 | 0.0322 | 0.39 | 0.22 | −43 | 3.3 |
| 40 | 0.244 | 0.77 | 0.57 | −26 | 4.3 |
| 50 | 0.343 | 1.12 | 0.816 | −27 | 3.4 |
| 60 | 0.481 | 2.73 | 0.664 | −62 | 5.0 |
| 6″ skirted cone (ref. 39a) | | | | | |
| 30 | 0.090 | 0.85 | 0.35 | −59 | 1 |
| 40 | 0.155 | 0.97 | 0.40 | −59 | 1 |
| 50 | 0.26 | 1.32 | 0.75 | −43 | 1 |
| 60 | 0.40 | 1.75 | 1.55 | −11 | 1 |
| 70 | 0.61 | 3.35 | 2.85 | −15 | 1 |

studying cavitation problems in closed conduits, the pressure and velocities can be controlled independently and it is therefore usually possible to obtain the proper $\sigma$ and therefore achieve similarity. There are, however, scale effects associated with the model being smaller in size and sometimes tested in a reduced velocity condition.

Past research has identified scale effects associated with variation in air content and water quality and the fact that the model is smaller and usually operated at reduced velocities and pressures. Most scale effects indicate that the prototype will cavitate more severely than predicted by the model. The size scale effects appear to be related to the length of time that the cavitation nuclei are subjected to the mean and fluctuating low pressures in separation regions. As the size increases, the nuclei are subjected to the low pressures for a longer period of time. This allows more of the nuclei to grow into cavitation events. Scale effects related to changes of velocity are likely due to a combination of variation of the time of the nuclei in the low-pressure region and the scaling of turbulence intensity with velocity.

When applying these same principles to scale effects for aeration, one would expect that the amount of air (per unit jet diameter) required to suppress cavitation should decrease as the size of the system increases and increase with velocity.

A research project was conducted to evaluate the amount of air required to suppress cavitation for orifices and evaluate scale effects (11). The results are summarized in Fig. 6.11. The air required to suppress cavitation is expressed in standard cubic feet per minute per foot of jet circumference (SCFM/jetcirc. ft). For data taken at three different jet velocities, the data show that the amount of air required (per foot of jet circumference) decreases with pipe diameter. For models 12 in. or larger, the amount of air becomes constant. For modeling, this shows that reduced size models

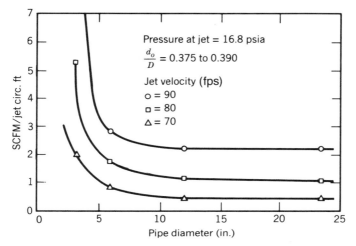

**Fig. 6.11**   Unit air flow versus pipe size.

(smaller than 12-in. dia.) will predict more air than will be required by the prototype. The data also show that the percent air by volume increases with velocity. Note that the total amount of air increases with pipe size, but the amount of air per foot of jet circumference decreases.

For estimates of the minimum amount of air needed, see Reference 11. If a system can tolerate large amounts of air, it would be advisable to provide more than the minimum. As a rough guide, for butterfly valves, the author has used two air vents placed in line with the shaft, as close as possible, each having a diameter 1/12 of the pipe diameter. This criterion has worked successfully for valves that have a short pipe downstream and free discharge into the atmosphere.

## PROBLEMS

**6.1.** During tests on a 914-mm butterfly valve, the following data were obtained: $P_{atm} = 82.8$ kPa, $P_{va} = 2.1$ kPa, $Q = 3.2$ m³/s, $f = .0133$, $P_u = 166$ kPa, and $P_d = 140$ kPa. $P_u$ and $P_d$ were measured 1 diameter upstream and 10 diameters downstream from the valve. Calculate the net pressure drop, $C_d$, and $\sigma$, and estimate the level of cavitation ignoring scale effects.

*Answer*:   $\Delta P = 24.3$ kPa, $C_d = 0.574$, $\sigma = 9.32$, no cavitation

**6.2.** Estimate the cavitation level for Problem 1 and include scale effects.

*Answer*:   light cavitation

**6.3.** A 305-mm globe valve will be installed where the downstream pressure will be constant at 305 kPa, $P_{vg} = -86$ kPa. Find the flow rate and allowable head drop for the valve 20% open so that it operates at critical cavitation. Use the type-A valve in Fig. 6.7 and include scale effects.

*Answer*: $\Delta P = 566$ kPa, $Q = 0.062$ m$^3$/s $(C_d = 0.025, \sigma_{co} = 0.60)$

**6.4.** A 6-in. butterfly valve is operating under the following conditions: $P_u = 69$ psi, $P_{vg} = -12.4 - \sigma_i$, $P_d = 42$ psi. Is the valve cavitating? $Q = 9.36$ cfs

*Answer*: near choking

**6.5.** A 12-in. type-A globe valve (Fig. 6.7) is operating at 100% open with $P_u = 75$ psi, $P_{vg} = -12.4$ psi, $P_d = 42$ psi. Is the valve cavitating? Calculate $Q$.

*Answer*: $Q = 11,800$ gpm

**6.6.** A 1.22-m diameter butterfly valve is installed and operated with $P_u = 640$ kPa, $P_{vg} = -90$ kPa, valve opening 60°. Find the allowable $P_d$ and discharge so the valve will operate at incipient damage.

*Answer*: $\sigma_{id} = 2.51$, $Q = 11.7$ m$^2$/s

**6.7.** A ball valve is tested to determine its critical cavitation index. The results were that at $P_{do} = 145$ kPa, $\sigma_{co} = 2.2$. What would be the value of $\sigma_c$ at $P_d = 250$ kPa, assume $P_{vg} = P_{vgo} = -80$ kPa?

*Answer*: $\sigma_c = 2.44$

**6.8.** Assume the following are laboratory data on a 76-mm butterfly valve at incipient cavitation: $P_{uo} = 620$ kPa, $P_{do} = 351$ kPa, $V_{io} = 4.22$ m/s, $P_{vgo} = -90$ kPa. Make no line loss corrections. Find: (a) $C_d$; (b) $\sigma_{io}$; (c) $\sigma_i$ at $P_d = 200$ kPa and $P_{vg} = -80$ kPa (d) $\sigma_i$ for a geometrically similar 305-mm butterfly valve operating at $P_d = 1000$ kPa and $P_{vg} = -84$ kPa and the percent error in $\sigma_i$ if scale effects are ignored.

**6.9.** A 36-in. butterfly valve is expected to be operating at critical avitation under the following conditions: $C_d = 0.267$, $P_{uo} = 12.7$ psi, $P_{do} = 7.70$ psi, $Q = 24,000$ gpm. If the valve is to be modeled with a 6-in. valve and tested at $P_u = 218$ psi what would be the model $\sigma_{co}$ at critical cavitation? (Assume $P_{vg} = -12.2$ psi.)

*Answer*: $\sigma_c = 5$, $Q = 3.92$ cfs

**6.10.** A 36-in. butterfly valve is installed to control flow from a reservoir. The valve is at an elevation of 2,000 ft and the reservoir level is at elevation 2,400 ft. Find the critical cavitation velocity when the valve is operating at 50% open. Also calculate the torque. Water $T = 50°$F. Use a type-A valve (Fig. 4.7).

*Answer*: $V = 20.3$ fps

# 7

# CAVITATION DATA FOR ORIFICES, NOZZLES AND ELBOWS

Valves are probably the most common flow control devices subject to cavitation because they are frequently used to create large pressure drops. However, there are situations where orifices or sudden enlargements, installed either for energy dissipation or flow measurement, are also operated in cavitation. Even elbows, tees, and other pipeline devices can be subjected to cavitation if the velocity is high or the pressure low. This chapter presents cavitation data for orifices, multihole orifices, nozzles, sudden enlargements, long-radius elbows, and wye-branches. The orifice data are most extensive because orifices were used as the primary test item to simulate simple valves and evaluate many of the scale effects. They are also commonly used for flow measurement.

## 7.1 ORIFICE SCALE EFFECTS

Orifice data were briefly discussed in Chapter 5, since they provided much of the information for size scale effects. Experimental data defining four cavitation levels for orifices installed in a 3-in. (76-mm) diameter pipeline operating at an upstream pressure of 90 psi (620 kPa) are summarized in Table 7.1 and Figs. 7.1 and 7.2. The incipient and critical data are from Reference 18, the damage data from References 59 and 60, and the choking data from Reference 63. Before using the data in Table 7.1, the reader is cautioned to review the scale effects discussed in Chapters 5 and 6 to account properly for the variation of these levels with discharge coefficient, pipe size, and operating pressure. This dependence varies with the type of control device and is different for the various cavitation levels. These scale effects are briefly reviewed.

**TABLE 7.1   Cavitation limits for orifices**[a]

| $d_o/D$ | $A_p/A_o$ | $A_p/A_j$ | $C_d$ | $\sigma_i$ | $\sigma_c$ | $\sigma_{id}$ | $\sigma_{ch}$ |
|---------|-----------|-----------|-------|------------|------------|---------------|---------------|
| 0.389 | 6.60 | 10.5 | 0.100 | 1.10 | 0.96 | 0.45 | 0.27 |
| 0.444 | 5.06 | 8.00 | 0.133 | 1.30 | 1.00 | 0.67 | 0.32 |
| 0.500 | 4.00 | 6.22 | 0.179 | 1.62 | 1.20 | 0.83 | 0.39 |
| 0.667 | 2.25 | 3.31 | 0.385 | 3.38 | 2.16 | 1.73 | 0.74 |
| 0.80C0 | 1.56 | 2.17 | 0.648 | 6.62 | 3.89 | 3.19 | 1.78 |

[a]Note: $D$ = 3-in., $P_{uo}$ = 90 psi, $P_{vgo}$ = −12.2 psi, $d_o$ = orifice diameter, $D$ = 3-in. pipe diameter, $A_p$ = pipe area, $A_o$ = orifice area, and $A_j$ = jet area.

### Incipient and Critical Cavitation

Extensive research has shown that there are no pressure scale effects associated with incipient and critical cavitation for orifices or sudden enlargements (7, 19). Therefore, the values of $\sigma_i$ and $\sigma_c$ in Fig. 7.1 and Table 7.1 apply at any pressure.

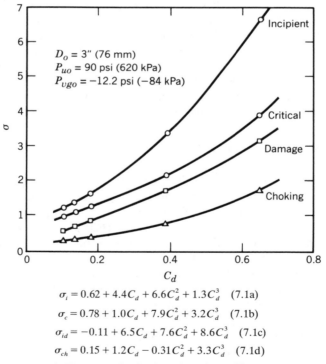

$$\sigma_i = 0.62 + 4.4C_d + 6.6C_d^2 + 1.3C_d^3 \quad (7.1a)$$

$$\sigma_c = 0.78 + 1.0C_d + 7.9C_d^2 + 3.2C_d^3 \quad (7.1b)$$

$$\sigma_{id} = -0.11 + 6.5C_d + 7.6C_d^2 + 8.6C_d^3 \quad (7.1c)$$

$$\sigma_{ch} = 0.15 + 1.2C_d - 0.31C_d^2 + 3.3C_d^3 \quad (7.1d)$$

**Fig. 7.1**   Cavitation data for orifices

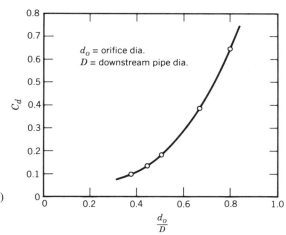

$$C_d = 0.019 + 0.083B - 0.203B^2 + 1.35B^3 \quad (7.1e)$$

$$B = d_o/D$$

$$B = 0.193 + 2.34C_d - 3.94C_d^2 + 2.73C_d^3 \quad (7.1f)$$

**Fig. 7.2**   Discharge coefficients for orifices.

The same research did uncover a measurable variation of $\sigma_i$ and $\sigma_c$ with pipe size. Both increase as the pipe diameter increases. The variation of $\sigma_i$ and $\sigma_c$ with size can be calculated using Eqs. 6.9–6.12 and the reference data in Table 7.1 or Fig. 7.1.

### Choking Cavitation

Studies on orifices and various types of valves made over a wide range of pressures and sizes have shown that there are no scale effects for choking cavitation. The value of $\sigma_{ch}$ is not affected by either size or pressure level.

### Incipient Damage

Size scale effects studies with geometrically similar orifices in 76-, 150-, and 300-mm pipes for a range of $d_o/D$ values did not show any variation of pitting rate with pipe size (59, 60). The value of $\sigma_{id}$ based on 1.0 pit/in.²/ min was the same for all three sizes. However, there was a significant increase in the size of the pits as the pipe size increased, just as was found for the tests at increasing pressures. Consequently, if incipient damage were based on volume or weight loss, there would be a size scale effect. If a system is operating at $\sigma < \sigma_{id}$, there will be more damage in larger systems. A simple relationship has not been developed to quantify this scale effect. See Reference 59 for more information.

The same study did show a measurable pressure scale effect. Adjusting $\sigma$ for pressure can be done with the same equation as used for valves:

$$\sigma_{id} = \sigma_{ido} \cdot \text{PSE} \quad (7.2)$$

with

$$\text{PSE} = \left[ \frac{(P_d - P_{vg})}{(P_{do} - P_{vgo})} \right]^{0.19} \tag{7.3}$$

or

$$\text{PSE} = \left[ \frac{(P_u - P_{vg})}{(P_{uo} - P_{vgo})} \right]^{0.19} \tag{7.4}$$

## 7.2   ZONES OF CAVITATION DAMAGE

If one plans to operate an orifice in damaging cavitation, it may be of interest to know where the damage will occur. Experimental results are available that identify the damage region (Fig. 7.3). Generally, the damage occurs farther from the orifice as the value of $d_o/D$ decreases. This is true for the upstream and downstream limits of pitting as well as the zone of maximum damage (Fig. 7.3). For simplification, the data points for the upstream and maximum pitting limits were not plotted: only the average lines were drawn. The upstream boundary of the pitting zone extended from $x/d_o = 0.4$ for $d_o/D = 0.80$ to $x/d_o = 1.3$ for $d_o/D = 0.389$ ($x = $ distance from orifice). The downstream boundary for the same $d_o/D$ ratios extended from about $x/d_o = 1.6$ to $x/d_o = 8.0$. The zone of maximum pitting also moved farther downstream as the orifice size decreased, varying from $x/d_o = 0.9$ to $x/d_o = 6.0$ over the same $d_o/D$ range.

These limits represented the results of 165 tests covering a range of cavitation intensities from incipient damage to near choking. The only

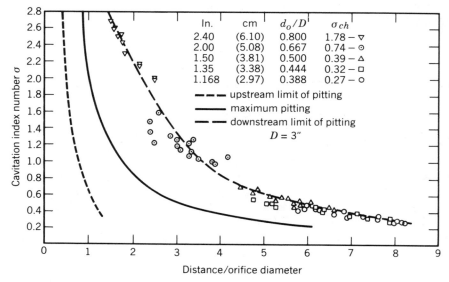

**Fig. 7.3**   Zone of pitting downstream from orifice.

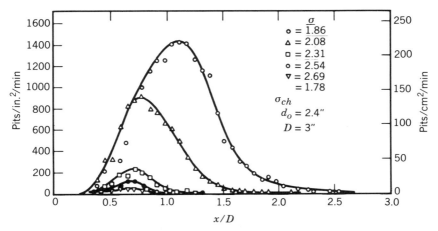

**Fig. 7.4**   Pitting distribution downstream from orifice.

condition that would cause the damage zone to exceed the identified limits is supercavitation.

### Pitting Distribution and Location

A typical pitting distribution in the damage area is shown in Fig. 7.4; the pits are few in number at the upstream limit, increase rapidly to a maximum and then decrease to the downstream limit. When pitting density was plotted versus distance, as shown in the figure, it was found that the upstream limit of cavitation damage for a particular orifice remained almost stationary for all levels of cavitation for which $\sigma > \sigma_{ch}$. The location of maximum pitting and the downstream limit moved downstream as the damage rate increased.

### Damage Rate

Traditionally, scale effects on damage have been evaluated by plotting damage rate versus velocity for a constant $\sigma$. Typical results indicate that damage rate is proportional to $V^n$, in which $V =$ the pipe velocity. Knapp (32, 33) determined values of $n$ to be in the range of four to six. Data for orifices in a 76-mm pipe from the orifice study gave values in the range of four to seven (59, 60). This indicates the strong dependence of damage on velocity and serves as a warning that once damage begins it can be increased significantly by a slight increase in velocity.

### Influence of Deaeration on Damage

During the cavitation damage tests on orifices, it was observed that when the downstream pressure was low and when the pressure drop was high, significant amounts of air came out of solution. The free air inhibits the formation of vaporous cavitation and also cushions the collapse of the cavities. It therefore reduces the chance of cavitation damage, and a lower $\sigma$ is required to cause damage.

This deaeration action has two interesting facets. First, the suppression of cavitation damage by two-phase air–water flow can provide an effective method of reducing cavitation damage if the mean pressure in the cavitation zone is subatmospheric (71). This is similar to injecting air to reduce damage (15, 40–72). Secondly, the data from damage tests made with subatmospheric pressures downstream should not be used for scaling to higher pressures. Serious errors would result at higher pressures where deaeration does not occur.

## 7.3   NOZZLES AND SUDDEN ENLARGEMENTS

The performance of nozzles and sudden enlargments can be predicted from orifice data. The incipient, moderate, and severe cavitation levels for orifices and nozzles were compared by Ball (5). He applied contraction coefficients for free flow to orifice data and plotted a cavitation index number for each of the three levels versus the jet-enlargement area ratios. He reasoned that the contraction coefficients for submerged orifice flow would not differ significantly from those for free flow. Good agreement was found between orifice and nozzle data.

Contraction coefficients $C_c$ for circular, concentric sharp-edged orifices can be found on page 57 of Reference 48. The values are for two-dimensional jets, but experimental data have verified their accuracy for three-dimensional flow. The data were fit to a polynomial equation valid for $0 < d_o/D < 0.8$.

$$C_c = 0.611 + 0.024\left(\frac{d_o}{D}\right) - 0.013\left(\frac{d_o}{D}\right)^2 + 0.196\left(\frac{d_o}{D}\right)^3 \quad (7.5)$$

Momentum principles were applied to orifice flow by Sweeney (60). He computed effective jet size or "vena contracta" to calculate contraction coefficients for orifices. He concluded that data could be applied to the design of sudden-enlargement energy dissipators formed by nozzles in conduits or smaller pipes discharging into larger pipes using whatever cavitation level is desired. One must be careful to make the comparison based on the jet area, not the orifice area. Scaling nozzle data would be done the same as for orifices.

Incipient cavitation for nozzles and pipe expansions computed by the preceding method may still deviate slightly from orifice data in view of work by Rouse and Jezdinsky (49). They showed that cavitation will not only be affected by the geometry of the jet, but also by the entrance conditions of the expansion. The intensity of the eddy formations and therefore the cavitating conditions are dependent on the velocity gradient around the jet as it enters the expansion. This gradient is steep for orifices, but becomes more gradual for nozzles or pipes discharging into sudden enlargements because a boundary layer develops. This results in a conservative prediction of cavitation levels for nozzles from orifice data.

## 7.4 MULTIHOLE ORIFICES

The concept of using a multiple-hole orifice plate device has been implemented in several applications for creating pressure reductions and reducing noise. A three-hole orifice was used in a large outlet tunnel from a dam in Canada (51). It is used in many of the newer-type high-pressure drop devices (80). It has been used also as a pressure-reducing device in circulating water systems of nuclear power plants (71). There are numerous other situations where the multihole orifice plate or its equivalent has been used in hydraulic systems. A good discussion of different valves that have used this concept is contained in Reference 80.

The multihole plate offers several advantages and some disadvantages. The most obvious disadvantage is that for it to be used in the system, the system must be free of any sizable suspended material. Some of the advantages are that it produces less noise and vibrations, it can be spaced close together in the event that orifices in series are required, and cavitation damage, if it occurs, is restricted to a shorter distance downstream from the orifice plate.

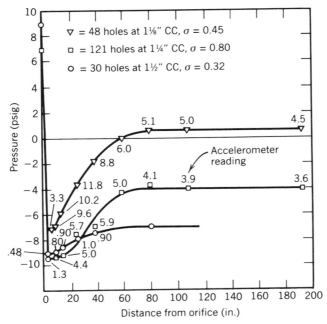

**Fig. 7.5** Pressure recovery for multihole orifice in 20-in. pipe.

Figure 7.5 shows the hydraulic grade line downstream from a multihole orifice plate. These data indicate that the recovered hydraulic grade line is established within 3 to 4 diameters. This suggests that such orifice plates can be spaced approximately 3 pipe diameters apart and still produce essentially 100% of their normal head loss.

Figure 7.6 shows $C_d$ data for multihole, orifices, single-hole orifices, and nozzles. If the area ratio is based on the area of the jet rather than the area of the hole, the discharge coefficient for multihole orifices, single-hole orifices, and nozzles are the same. The fact that the multihole orifice discharge coefficient appears not to be a function of the number, the spacing, or diameter of the holes makes it possible to use single-hole orifice data to design multihole orifice plates.

Only limited data are available identifying the cavitation characteristics of multihole orifice plates. Reference 71 describes a study carried out with multihole orifice plates having a negative pressure downstream. For these tests, large amounts of air came out of solution because of the low downstream pressure and high turbulence generated by the jets. This air suppressed cavitation and prevented cavitation damage. It is anticipated that a multihole orifice plate operating at positive pressures will have cavitation characteristics slightly worse than those of a single-hole orifice plate. The one difference that has been observed is that when damage occurs, it only occurs close to the orifice, rather than extending for several pipe diameters downstream.

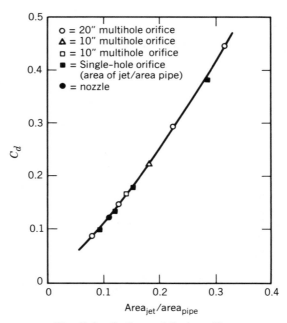

**Fig. 7.6** $C_d$ for multihole orifice.

Reference 46 studied damage caused by a multihole orifice. The orifice had 16 1/2-in. diameter sharp-edged holes with an equivalent orifice area of 3.14 in.$^2$. It was mounted in a 6-in. diameter pipe and had a measured $C_d = 0.075$. Tests were conducted at $P_u = 90$ psi and incipient damage occurred at $\sigma_{id} = 0.41$. The test results were compared to a $6 \times 2$ in. orifice, $C_d = 0.071$ and $\sigma_{id} = 0.30$. It was suggested that damage occurred at a larger $\sigma$ for the multihole orifice because the outside row of holes is closer to the pipe wall than the single jet. Damage was observed to occur closer to the orifice for the multihole plate.

For orifices (or valves) placed in series, the combined discharge coefficient can be calculated by the equation

$$ C_d = \left[ \frac{1}{1 + \dfrac{1}{C_{d1}^2} + \dfrac{1}{C_{d2}^2} \cdots \dfrac{1}{C_{dn}^2} - N} \right]^{0.5} \tag{7.6} $$

in which $C_{d1}$ is the discharge coefficient for the first orifice plate, $C_{d2}$ for the second orifice plate, etc., and $N$ is the total number of orifices.

Three different combinations of three plates placed in series were calibrated to determine if the combined $C_d$ of each set of plates was affected by hydraulic interference caused by the close spacing. The orifice plates were installed 2.5 pipe diameters apart. The three sets of data showed that for a spacing of 2.5 diameters, the combined discharge coefficient was approximately 5% larger than calculated. This was because the pressure was not completely recovered, as shown in Fig. 7.5, within the 2.5 pipe diameters. If desired, the proper $C_d$ could be obtained by reducing the number of holes.

## 7.5 LONG RADIUS ELBOWS

Elbows and other pipe components are far less likely to cavitate than valves, pumps, or orifices. Nevertheless, they can cavitate and there are situations in which it is important to be able to predict the cavitation intensity. Incipient and critical cavitation data are listed in Table 7.2 for a 6-in. commercial long-radius elbow. The cavitation index used for the elbow was

$$ \sigma_3 = 2(P_t - P_{vg})/\rho V^2 \tag{7.7} $$

in which $P_t$ was the total upstream pressure (gauge pressure plus velocity head pressure). The reason for selecting $\sigma_3$ was that for the model study from which the data were obtained, it was not possible to measure a pressure drop for the elbow. The choice of using $P_t$ was also related to the experimental setup. This form of the equation is preferable to one using a pressure drop because there is considerable variation in published values of $K_l$ for elbows.

**TABLE 7.2   Values of Incipient and Critical Cavitation Index for a 6-in. Long-Radius Elbow**

| Test condition (1) | | $\sigma_{3i}$ (2) | $\sigma_{3c}$ (3) |
|---|---|---|---|
| *Constant Velocity* | | | |
| 18.3 m/s | | 3.40 | 2.40 |
| 15.2 m/s | | 3.60 | 2.70 |
| 12.2 m/s | | 3.60 | 2.70 |
| | Average: | 3.53 | 2.60 |
| *Constant Local Upstream Pressure Head* | | | |
| 97.3 kPa | | 3.30 | 2.75 |
| 172.0 kPa | | 3.20 | 2.85 |
| 276.0 kPa | | 3.15 | 2.50 |
| 414.0 kPa | | 3.40 | 2.50 |
| | Average: | 3.26 | 2.65 |
| *Constant Upstream Gauge Pressure* | | | |
| 34.5 kPa | | 3.35 | 2.75 |
| 104.0 kPa | | 3.20 | 2.80 |
| 172.0 kPa | | 3.35 | 2.70 |
| | Average: | 3.30 | 2.77 |

**TABLE 7.3   Scale Effect Results for Long-Radius Elbows**[a]

| Size (cm) (1) | $P_u$ (kPa) (2) | $\sigma_{3i}$ (3) | $\sigma_{3c}$ (4) | $\sigma_{3ch}$ (5) | $\sigma_{3id}$ (6) |
|---|---|---|---|---|---|
| 7.62 | 97.3 | 2.40 | 2.25 | 1.88 | |
| | 172.0 | 2.40 | 2.20 | 1.88 | |
| | 276.0 | 2.40 | 2.25 | 1.88 | |
| | 414.0 | 2.40 | 2.30 | 1.92 | |
| 15.2 | 97.3 | 3.30 | 2.75 | 1.87 | |
| | 172.0 | 3.20 | 2.85 | 1.90 | |
| | 276.0 | 3.15 | 2.50 | 1.93 | |
| | 414.0 | 3.40 | 2.50 | 1.94 | |
| 30.5 | 97.3 | 4.60 | 4.00 | 1.88 | 2.2 |
| | 172.0 | 4.60 | 4.00 | | |
| | 276.0 | 4.50 | 3.70 | | |

$$^a\sigma_3 = \frac{2(P_t - P_{vg})}{\rho V^2}.$$

176

The data in Tables 7.2 and 7.3 were taken at constant velocity, constant upstream gauge pressure, and constant total upstream pressure. The reason for the three sets of tests was to determine if the method of testing influenced the scale effects. The data show that there are no velocity or pressure scale effects for the elbow. Table 7.3 shows a significant size scale effect for $\sigma_{3i}$ and $\sigma_{3c}$, but none for choking.

After several months of testing on the model elbow, it was removed and inspected for zones of cavitation damage. The interior of the elbow had been painted to protect against corrosion. Significant paint removal was observed on the inner curve of the elbow approximately 90° from the inlet. Some of the paint was definitely removed by cavitation, since pitting of the exposed steel was also observed.

To determine the incipient damage condition for the elbow and investigate pressure scale effects, aluminum buttons were installed in the zone where cavitation damage was observed. Cavitation damage tests were conducted only with the 305-mm diameter elbow. Tests were carried out at three values of constant total upstream pressure, 34.5, 69.0, and 97.0 kPa. The variation of damage rate as a function of $\sigma_3$ was plotted on logarithmic paper and a linear variation in damage with increasing $\sigma$ was observed. At the defined incipient damage point of 1 pit/in.$^2$/min on soft aluminum, the incipient damage condition was $\sigma_{3id} = 2.2$ for all pressures, indicating that there is no pressure scale effect.

As previously discussed, orifices and nozzles have shown a pressure scale effect on $\sigma_{id}$. The work on the orifices and nozzles was more extensive and carried out over a wider range of conditions. It is therefore possible that if a continuous strip of aluminum could have been used in the elbow so the exact location of inception could have been determined, and if the range of conditions had been increased, a pressure scale effect might have been found for the elbow.

A summary of the values of the choking cavitation index for the elbows at different upstream total heads is in Table 7.3. No consistent variation of $\sigma_{3ch}$ with size or pressure is observed. It is seen that there is less than 2% variation from the mean value of 1.9, indicating no scale effect.

## 7.6  COMMENTS ON SIZE SCALE EFFECTS

In applying any of the size scale effect equations to incipient or critical cavitation for elbows, orifices, or valves, it is important to realize that the equations are inherently conservative. This is because of the way incipient and critical cavitation are determined. They are determined based on an absolute number of cavitation events rather than on events per unit area or per unit volume of flow.

Dynamic similarity would imply that the same percentage of nuclei would be cavitating in systems of different size. As a result, if one compared the cavitation conditions for a 76- and a 305-mm pipe with the same percentage of nuclei per unit volume of water growing into vapor cavities, and the velocity of flow were the same in both pipes, the larger system would be expected to have 16 times as many cavitation events per unit of time as the small one. Incipient or critical as defined in Chapter 5 requires both systems to operate at approximately the same number of events per unit of time. This forces the larger system to operate at a larger value of the cavitation index. This means that the incipient and critical cavitation conditions for differently sized pipes will sound about the same intensity, but for a larger system the potential for vibration and erosion damage is less because there are fewer events per unit volume of flow.

It is important to realize that this does not account for all of the size scale effects. Larger systems operating at the same $\sigma$ definitely produce more cavitation as well as a more severe cavitation.

**Example 7.1.** An orifice with $d_o/D = 0.47$ is to be installed in a 76-mm pipe where $P_u = 680$ kPa, and $P_{vg} = -84$ kPa. Calculate the allowable pressure drop if the orifice is to operate at (a) critical cavitation and (b) incipient damage.

*Case a. Critical cavitation.* Using Fig. 7.2, find $C_d = 0.153$, and from Fig. 7.1 get $\sigma_c = 1.12$ (remember that there are no pressure scale effects). Calculate $\Delta P$ allowable from:

$$P_d = \frac{(\sigma_c P_u + P_{vg})}{(1 + \sigma_c)} = 320 \text{ kPa}$$

so

$$\Delta P = 680 - 320 = 360 \text{ kPa at critical cavitation}$$

*Case b. Incipient Damage.* Using the same procedure as in Case a:

$$P_{uo} = 620 \text{ kPa}, \ C_d = 0.153, \ X = 0.19 \text{ (Table 6.2)}$$

$$P_{vo} = -84 \text{ kPa}, \ \sigma_{ido} = 0.75$$

Adjust for pressure scale effects:

$$\sigma_{id} = 0.75[(680 + 84)/(620 + 84)]^{0.19} = 0.76$$

$$P_d = (\sigma_{id} P_u + P_{vg})/(1 + \sigma_{id}) = 246 \text{ kPa}$$

$$\Delta P = 680 - 246 = 434 \text{ kPa}$$

*Example 7.2.* An orifice with a hole diameter of 9.84 in. is installed in a 16-in. pipe. It will operate at a constant upstream pressure of 75.4 psi, $P_{va} = 1.74$ psi, and $P_b = 13.9$ psi. Find $\sigma_i$.

Use the data in Figs. 7.1 and 7.2 and adjust for size scale effects:

$$d_o/D = 9.84/16 = 0.615, \quad C_d = 0.305 \text{ (Fig. 7.2)}, \quad K_l = 9.75$$

$$\sigma_{io} = 2.55 \text{ (for } D = 3\text{-in. pipe, Fig. 7.1)}$$

Use Equation 6.9–6.11 for size scale effects ($D = 16$, $d = 3$):

$$Y = 0.3(9.75)^{-0.25} = 0.167$$

$$\text{SSE} = 5.33^{0.167} = 1.32$$

$$\sigma_i = 1.32 \cdot 2.55 = 3.37$$

## Valve and Orifice in Series

If large pressure drops are required and the range of flow is not large, it may be possible to use one valve plus one or more orifice plates in series rather than using valves in series, as done in Example 6.5. The obvious advantage to the orifice plates is that they are much less expensive. The disadvantage is that they are a fixed resistance and all control of flow must be provided by the valve. If a wide range of flow control is required, it is likely that a single valve with orifice plates will result in the valve cavitation at smaller flow rates.

*Example 7.3.* The problem in Example 6.6 will be redesigned using one globe valve and one or more orifice plates in series with all devices operating near incipient damage.

*Design principles*:

1. To provide the maximum controllability, the system should be designed so that the valve is full open at maximum discharge. Therefore, start the design by selecting orifice plates that provide the proper discharge and operate at or below incipient damage (or whatever level is desired).

2. The design procedure is trial and error because it is necessary to first select a trial orifice diameter and then compare the allowable pressure drop, based on cavitation, with the actual pressure drop at the given pipe velocity.

**TABLE 7.4  Spreadsheet for Example 7.3**

| | A | B | C | D | E | F | G | H | I | J | K | L | M | N | O | P | Q | R | S | T | U | V |
|---|---|---|---|---|---|---|---|---|---|---|---|---|---|---|---|---|---|---|---|---|---|---|
| 1 | DEVICE | Flow | ID | $P_u$ | $P_d$ | V | $P_{ug}$ | $\sigma$ | $P_{ugo}$ | $C_d$ | $\sigma_{id}$ | $P_{uo}$ | $P_{do}$ | SSE | X | PSE | $\sigma_{id}$ | $d_o/D$ | $d_o$ | $A_o$ | Hold Dia. | # Holes |
| 2 | | cfs | in. | psi | psi | fps | psi | sys. | psi | | Eqn. | psi | psi | | | | adj. | Eqn. | in. | in. ×2 | in. | |
| 3 | #1 orifice | 11.7 | 12.0 | 25.0 | 0.0 | 14.9 | −12.9 | 0.52 | −12.2 | 0.237 | 1.10 | 90.0 | 41.3 | 1.00 | 0.19 | 0.83 | **0.91** | 0.56 | 6.76 | 36 | 1.5 | 20.3 |
| 4 | #1 orifice | 11.7 | 12.0 | 10.0 | 0.0 | 14.9 | −12.9 | 1.29 | −12.2 | 0.360 | 1.63 | 90.0 | 51.1 | 1.00 | 0.19 | 0.75 | **1.22** | 0.65 | 7.83 | 48 | 1.5 | 27.2 |
| 5 | | | | | | | | | | | | | | | | | | | | | | |
| 6 | #2 orifice | 11.7 | 12.0 | 25.0 | 10.0 | 14.9 | −12.9 | 1.53 | −12.2 | 0.301 | 1.37 | 90.0 | 46.9 | 1.00 | 0.19 | 0.83 | **1.13** | 0.61 | 7.38 | 43 | 1.5 | 24.2 |
| 7 | #2 orifice | 11.7 | 12.0 | 31.0 | 10.0 | 14.9 | −12.9 | 1.09 | −12.2 | 0.258 | 1.18 | 90.0 | 43.2 | 1.00 | 0.19 | 0.85 | **1.01** | 0.58 | 6.97 | 38 | 1.5 | 21.6 |
| 8 | | | | | | | | | | | | | | | | | | | | | | |
| 9 | #3 orifice | 11.7 | 12.0 | 62.0 | 31.0 | 14.9 | −12.9 | 1.42 | −12.2 | 0.214 | 1.00 | 90.0 | 38.8 | 1.00 | 0.19 | 0.94 | **0.94** | 0.54 | 6.49 | 33 | 1.5 | 18.7 |
| 10 | | | | | | | | | | | | | | | | | | | | | | |
| 11 | 12-in. globe | 11.7 | 12.0 | 71.0 | 62.0 | 14.9 | −12.9 | 8.32 | −12.2 | 0.377 | 1.59 | 90.0 | 50.5 | 1.00 | 0.30 | 0.94 | **1.50** | | | | | |
| 12 | | | | | | | | | | | | | | | | | | | | | | |
| 13 | #1 orifice | 5.00 | 12.0 | 1.8 | 0.0 | 6.37 | −12.9 | 7.04 | −12.2 | 0.360 | 1.62 | 90.0 | 51.0 | 1.00 | 0.19 | 0.69 | **1.12** | 0.65 | 7.83 | 48 | 1.5 | 27.2 |
| 14 | #2 orifice | 5.00 | 12.0 | 3.8 | 1.8 | 6.37 | −12.9 | 7.39 | −12.2 | 0.258 | 1.19 | 90.0 | 43.2 | 1.00 | 0.19 | 0.71 | **0.84** | 0.58 | 6.98 | 38 | 1.5 | 21.6 |
| 15 | #1 orifice | 5.00 | 12.0 | 5.7 | 3.8 | 6.37 | −12.9 | 9.00 | −12.2 | 0.214 | 0.99 | 90.0 | 38.8 | 1.00 | 0.19 | 0.72 | **0.72** | 0.54 | 6.48 | 33 | 1.5 | 18.7 |
| 16 | 12-in. globe | 5.00 | 12.0 | 71.0 | 5.7 | 6.37 | −12.9 | 0.28 | −12.2 | 0.064 | 0.29 | 90.0 | 10.6 | 1.00 | 0.30 | 0.94 | **0.27** | | | | | |
| 17 | | | | | | | | | | | | | | | | | | | | | | |
| 18 | Column | Source of data | | | | | | | | | | | | | | | | | | | | |
| 19 | A | Input | | | | | | | | | | | | | | | | | | | | |

| | | |
|---|---|---|
| 20 | B | Given data |
| 21 | C | Calculated during preliminary calculations |
| 22 | D | Assumed |
| 23 | E | Initially given as 0, for upstream valves it equals $P_u$ of downstream valve |
| 24 | F | $Q$/area |
| 25 | G | Calculated during preliminary calculations |
| 26 | H | Calculated system $\sigma = (P_d - P_{vg})/(P_u - P_d)$ |
| 27 | I | Found from reference data (Table 7.1 for this example) |
| 28 | J | $V/(2g(P_u - P_d)) \cdot 2.308 + V \times 2)^{0.5}$ |
| 29 | K | Reference cavitation data (Eq. 7.1c used for this example) |
| 30 | L | Found from reference data (Table 7.1, for this example) |
| 31 | M | $[\sigma(ido) \cdot P_{uo} + P_{vgo})/(1 + \sigma(ido)]$ |
| 32 | N | Normally use Eq. 6.9, but since there are no SSE for incipient damage, enter 1.0 |
| 33 | O | From Table 6.1 or 6.2 |
| 34 | P | Calculated with Eq. 6.7b for PSE |
| 35 | Q | Calculated with Eq. 6.12 for PSE and SSE |
| 36 | R | Eq. 7.1f |
| 37 | S | $d_o/D \cdot D$ |
| 38 | T | $0.7854 \cdot d_o \times 2$ |
| 39 | U | Assumed |
| 40 | V | $A_o/(0.7854 \cdot \text{hole dia.} \times 2)$ |

*Orifice #1* (downstream). Keeping the diameter at 12 in. and referring to Example 6.6 and Fig. 7.1, the basic data for the first orifice are

$$V = 14.9 \text{ fps}$$

$$P_d = 0$$

$$P_{vgo} = -12.2 \text{ psi (Table 7.1)}$$

$$P_{uo} = 90 \text{ psi (Table 7.1)}$$

$$P_{vg} = -12.9 \text{ psi}$$

$$P_u = 25 \text{ psi (assumed)}$$

$$C_d = 14.9/(64.4 \cdot 2.308 \cdot 25 + 14.9^2)^{0.5} = 0.237$$

$$X = 0.19 \text{ (Table 6.2)}$$

$$\sigma_{ido} = 1.10 \text{ (Eq. 7.1c)}$$

$$P_{do} = \frac{(\sigma_{ido}P_{uo} + P_{vgo})}{(1 + \sigma_{ido})} = 41.3 \text{ psi}$$

$$\sigma_{system} = (0 + 12.9)/25 = 0.52$$

Adjust for pressure scale effects.

$$\sigma_{id} = 1.10 \left[ \frac{(0 + 12.9)}{(41.3 + 12.2)} \right]^{0.19} = 0.91$$

Since $\sigma_{system} < \sigma_{id}$, the orifice is too small. The calculations must be repeated again assuming $P_u < 25$ psi. Table 7.4 is the output of a spreadsheet to do these calculations. Note that row 3 contains the calculations for the first estimate. The $P_u$ in column D must be reduced until the $\sigma_{system}$ in column H is greater than the adjusted $\sigma_{id}$ in column Q. The first orifice can only produce a pressure drop of 10 psi and operate below incipient cavitation.

For the next orifice, $P_d = P_u$ for the downstream orifice. The calculations are continued until enough orifices are selected to drop most of the available pressure of 71 psi. Note that the 12-in. globe valve will cause a pressure drop of 9 psi when it is wide open. This leaves 62 psi for the orifices. For this example, it takes three orifices to prevent damaging cavitation. Since the third orifice is operating well away from damage, it would be advisable to reduce the pressure drop across the other orifices so they are all about the same distance from damage. The final choice of the orifices for this example will be those in rows 4, 7 and 9.

The last five columns show the size of a single hole orifice (col. S) and the number of holes of 1.5-in. diameter if it were desired to use a multihole orifice. For example, orifice #1 would either be a single-hole plate with a 7.83-in. diameter hole or a multihole plate with 27 holes 1.5-in. diameter.

The analysis should be repeated at smaller flow rates to detect when the globe valve begins to cavitate. This is done by keeping the orifice sizes fixed and calculating the pressure drop across them at reduced flow. The spreadsheet must be modified slightly. Rows 13-16 show the calculations. The values of $C_d$ for the orifices are fixed at the values in rows 4, 7 and 9. The $P_u$ values are calculated by

$$P_u = P_d + \left( \frac{1}{C_d^2} - 1 \right) \cdot \frac{V^2}{(2.308 \cdot 2g)}$$

At 5.00 cfs, the three orifices only produce 5.7-psi pressure drop. The valve drops 65.3 psi and operates at $\sigma_{system}$ which is just larger than $\sigma_{id}$. Therefore, the valve can control the flow between 5.00 and 11.7 cfs without cavitation damage.

# 8

# FUNDAMENTALS OF HYDRAULIC TRANSIENTS

Designing a pipeline to operate under steady-state conditions involves application of well-established and fairly easy-to-understand principles. If assumptions such as pipe friction, minor losses, or aging factors are not exact, the consequence is usually that the actual flow is just slightly different from the calculated flow. The serious engineering problems with pipelines are usually associated with cavitation or unsteady flow conditions: either during filling the line, while making intentional changes or changes caused by power failure to pumps, accidental opening or closing of a valve, etc. For some pipelines, especially long lines over hilly terrain, the most critical time in the life of the pipe can be the initial filling, pressurizing, and flushing out of the air. If adequate design provisions and operational procedures are not provided, it may not be possible to safely place the line into service. Transient pressure generated during filling can easily exceed the safe operating range of the pipe.

Analyzing a system to anticipate the type and magnitudes of possible hydraulic transients is more complicated than performing the steady-state calculations. Since analysis of unsteady flow is a relatively new field, some practicing engineers have not had adequate exposure to the topic and totally ignore the problem. As a result, too many pipelines fail.

The purpose of Chapters 8–10 is to introduce the engineer to transients, give guidance on recognizing potential problems, indicate various means of controlling the problem, and offer methods of predicting the magnitude of the pressures. The treatment concentrates on transients caused by valve operation. Transients caused by pumps and turbines are treated elsewhere (87).

## 8.1 DEFINITIONS

The term transient refers to any unsteady flow condition. It can refer to a situation where conditions are continually varying with time or to transition flow between two steady-state conditions, the latter being most common. Examples would be changing the setting of a valve or turning on a pump. The larger the incremental change and the faster that change takes place, the greater the transient pressure. If the piping system is not designed to withstand the high transient pressures or if controls are not included to limit the amount of head rise, rupture of the pipe or damage to equipment and machinery can result. The literature contains examples of pipes that have been ruptured or collapsed, pipe anchors that have been broken, and pumps, turbines, valves, etc., that have been damaged by hydraulic transients.

Since transients can occur in gases as well as liquids, the term fluid transient is used to encompass all situations. This book deals only with liquids, so the term hydraulic transient is used. When referring to water, the most common term used to describe unsteady flow is waterhammer. This descriptive name is apt, since the noise associated with transients in small metal pipes often sounds like someone hitting the pipe with a hammer.

Transients involving changes that occur slowly are called surges. Examples would be an oscillating U-tube, establishment of flow after a valve is opened, and the rise and fall of the water level in a surge tank. The method of surge analysis, called "rigid column theory," usually involves mathematical or numerical solution of simple ordinary differential equations. The compressibility of the fluid and the elasticity of the conduit are ignored and the entire column of fluid is assumed to move as a rigid body.

When changes in velocity, and consequently pressure, occur rapidly, both the compressibility of the liquid and the elasticity of the pipe must be included in the analysis. This procedure is often called "elastic" or "waterhammer" analysis and involves acoustic pressure waves traveling through the pipe and the solution of partial differential equations. Even though the term transient refers to all unsteady flows, it is generally used to identify the "elastic" case specifically.

### Causes of Transients

Numerous events can generate a transient or surge. Some of the most common are:

1. Change of valve opening.
2. Starting or stopping pumps.
3. Operation of check valves, air-release valves, pressure-reducing valves, and pressure-relief valves.

4. Pipe rupture.
5. Improper filling, flushing, or removal of air from pipelines.
6. Trapped air in pipelines.
7. Change in power demand of hydraulic turbines.

All pipelines experience transients. Whether the transient creates operational problems or pipe failure depends on its magnitude and the ability of the pipes to tolerate high pressures without damage. For example, an unreinforced concrete pipeline may have a transient pressure allowance of only a few feet above its operating pressure before damage can occur. For such situations, even slow closing of control valves or minor interruptions of flow due to any cause may create sufficient transient pressures to rupture the pipeline. In contrast, steel pipelines can usually take relatively high transient pressures without failure.

It is recommended that every pipe system should have at least a cursory transient analysis performed to identify the possibility of serious transients and determine whether a detailed analysis is necessary.

### Methods of Controlling Transients

If an analysis indicates that transients may pose a problem, the types of solutions available to the engineer include:

1. Increase the closing and opening time of control valves.
2. Design special facilities for filling, flushing, and removing air from pipelines.
3. Increase the pressure class of the pipeline.
4. Limit the pipe velocity.
5. Reduce the wave speed by changing the type of pipe or injecting air.
6. Use of pressure relief valves, surge tanks, air chambers, etc.

This treatise should not be construed to be a state-of-the-art summary of all current knowledge and techniques for analyzing hydraulic transients, but is intended to: 1) provide a basic understanding of the cause of transients and their effects on the system, 2) discuss alternative means of controlling or preventing undesirable transients, 3) introduce hand computational techniques for simplified piping systems, 4) present the method of characteristics which is suitable for making detailed accurate calculations of simple or complex systems, 5) develop various boundary conditions, and 6) discuss problems associated with column separation and trapped air.

The rest of this chapter develops the relationship that predicts the transient head rise caused by a sudden change in fluid velocity, derives the equation for wave speed in an elastic conduit, discusses the sequence of events during a simple transient, quantifies the influence of undissolved air on the wave speed and transient pressures, and gives examples of simple transient calculations.

## 8.2 BASIC TRANSIENT EQUATION

An equation predicting the head rise $\Delta H$ caused by a sudden change of velocity $\Delta V = V_2 - V_1$ can be derived by applying the unsteady momentum equation to a control volume of a section of a pipe where the change of flow is occurring. Consider the straight horizontal frictionless pipe between a reservoir and control valve shown in Fig. 8.1$a$. Ignoring friction, the steady-state hydraulic grade line is horizontal and the entire reservoir head HR is dissipated by the valve. The assumption of a horizontal pipe and no friction simplifies understanding of the pressure waves, but does not place limitations on the following derivation.

Consider the case of a partial valve closure which instantly reduces the velocity by an amount $\Delta V$. Reduction of the velocity can only be accomplished by an increase in the pressure head upstream from the valve. This creates a pressure wave of magnitude $\Delta H$, which travels up the pipe at the acoustic velocity $a$ as shown in Fig. 8.1$a$. The increased pressure compresses the liquid and expands the pipe.

Figure 8.1$b$ shows the unsteady momentum equation applied to a control volume of a section of pipe through which the pressure wave is passing. The

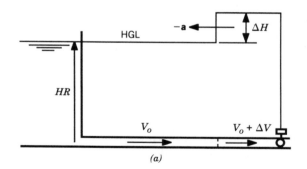

(a)

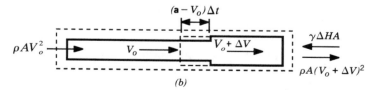

(b)

**Fig. 8.1**   Control volume for valve closure.

expansion of the pipe and the increase in fluid density are small compared with other changes and can be ignored. The net force acting on the control volume in the positive direction of flow is $-\gamma\Delta HA$. The net momentum flux through the control volume is $\rho A(V_o + \Delta V)^2 - \rho A V_o^2$, which reduces to $2\rho A V_o \Delta V$ by ignoring the $\Delta V^2$ term.

In a time $\Delta t$, the pressure wave traveling at an absolute velocity of $a\text{-}V_o$ moves a distance $(a\text{-}V_o)\Delta t$. The momentum within that length of pipe has been reduced because of the reduction in velocity. The mass of fluid affected is $\rho A(a - V_o)\Delta t$. The time rate of change of momentum within the control volume is the product of $\Delta V$ and the mass of fluid whose velocity is changed in $\Delta t$ seconds $\rho A(a - V_o)\Delta t\ \Delta V/\Delta t$ or $\rho A(a - V_o)\Delta V$.

Substituting these various terms into the momentum equation gives:

$$-\gamma\Delta HA = 2\rho A V_o \Delta V + \rho A(a - V_o)\Delta V$$

or

$$\Delta H = -\frac{a\Delta V}{g}\left(1 + \frac{V_o}{a}\right) \tag{8.1}$$

Since for many applications the wave speed is about 100 times the pipe velocity, the term $V_o/a$ can be dropped from Eq. 8.1. With this limitation the transient head rise due to an incremental change in velocity is:

$$\Delta H = -a\ \Delta V/g, \qquad \text{for } a \gg V \tag{8.2}$$

This equation is easy to use for multiple incremental changes of velocity as long as the first wave has not been reflected back to the point of origin. Equation 8.2 can therefore be expressed as

$$\Sigma\Delta H = \frac{-\Sigma a\Delta V}{g}, \qquad \text{for } a \gg V \tag{8.3}$$

***Example 8.1.*** Water is flowing at $V_o = 2$ m/s in a steel pipe. If a valve at the end of the pipe is partially closed so that the velocity is instantly reduced to 1.5 m/s, what is the transient head rise? Assume that the wave speed for water in the pipe is 900 m/s and that $g = 9.81$ m/s$^2$:

$$\Delta H = \frac{-900(-0.5)}{9.81} = 45.9 \text{ m}$$

It is apparent from this example that even small changes in the velocity cause large transient pressures. It is this simple principle that is responsible for transients playing such a vital role in the successful operation of pipelines.

## 8.3    PRESSURE WAVE PROPAGATION WITHOUT FRICTION

One limitation on the application of Eqs. 8.2 and 8.3 is that they are easy to use only until a reflected pressure wave returns to the location where $\Delta V$ was generated. To explain the propagation and reflection of the pressure wave, the sequence of events caused by a valve closure at the end of a pipe connected to a reservoir will be discussed. To simplify the problem, the pipe is assumed to have a constant diameter, be made of one material and pressure rating so it has a constant wave speed, and have no minor losses that would cause losses or reflections; friction is assumed to be negligible. Cases involving friction can be handled, but the analysis is slightly more complex. Figure 8.2$a$ shows the steady-state hydraulic grade line (HGL) for a pipeline of length having an initial velocity $V_o$ at $t = 0$.

If the valve is closed instantly at $t = 0$, the head rise $\Delta H$ is calculated using Eq. 8.2 with $\Delta V = -V_o$. The liquid immediately upstream from the valve comes to rest and the pressure increase causes both the liquid density and the pipe diameter to increase slightly. The pressure wave travels upstream at the acoustic wave speed $a$. Behind the wave, the hydraulic grade line is increased by $\Delta H$ and the liquid has zero velocity. Ahead of the wave, the velocity and pressure are at the initial steady-state values. The mass of liquid entering the pipe as the wave travels to the reservoir is just equal to the increased volume of the pipe due to expansion plus the added mass stored due to increased liquid density. Figure 8.2$b$ shows conditions in the pipe at $t = 0.75L/a$.

At $t = L/a$ the wave arrives at the reservoir. The pressure inside the pipe is $HR + \Delta H$, the velocity is everywhere zero, and the pipe is expanded and the fluid compressed. In the reservoir at the entrance to the pipe the pressure is still $HR$. The system is in nonequilibrium and the liquid in the pipe acts like a compressed spring, causing flow from the pipe back into the reservoir at the initial velocity $-V_o$. The reverse velocity is equal to the initial velocity because friction losses are being ignored and Eq. 8.2 still applies.

A negative pressure wave of magnitude $-\Delta H$ travels toward the valve at speed $a$. This causes the pressure in the pipe behind the wave to drop to $HR$ and the velocity to $-V_o$. The pipe diameter and the fluid density return to their original values. In front of the wave, the head stays constant at $HR + \Delta H$ and the velocity is zero. Figure 8.2$c$ shows conditions at $t = 1.5\ L/a$.

When the wave arrives at the valve at $t = 2L/a$ for an instant, the velocity everywhere is at $-V_o$, the head is at $HR$, and the pipe diameter and density are at their original values. The mass of fluid discharging into the reservoir between $t = L/a$ and $2L/a$ is the amount that entered the pipe during the first $L/a$ seconds and was stored in liquid compression and pipe expansion.

At $t = 2L/a$, the liquid at the valve tries to accelerate toward the reservoir at velocity $-V_o$. Since the valve is closed, and ignoring the possibility of the

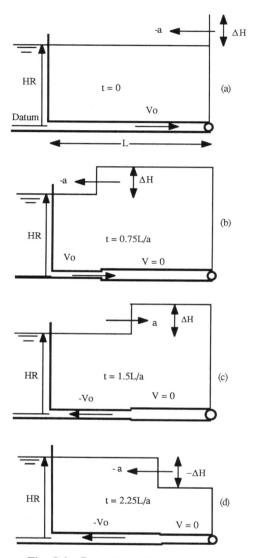

**Fig. 8.2**   Pressure-wave propagation.

liquid column rupturing due to vaporization, the velocity at the closed valve must be zero at all times. In preventing the fluid from accelerating to $-V_o$, the valve exerts a tension force that causes the head to drop by $-\Delta H$, as predicted from Eq. 8.2. This negative wave travels toward the reservoir at the acoustic velocity. Behind the wave, the pressure is at $HR - \Delta H$, the pipe diameter reduces, and the density decreases slightly. Ahead of the wave, the liquid continues to move at $-V_o$ into the reservoir. Figure 8.2$d$ shows conditions at $t = 2.25L/a$.

At $t = 3L/a$, the wave arrives at the reservoir. The pressure in the pipe is $-\Delta H$ below the reservoir pressure and the velocity is momentarily everywhere zero. The pressure imbalance causes flow to enter the pipe at $V_o$ and returns the hydraulic grade line to $HR$. As this positive pressure wave of $\Delta H$ travels to the valve, the velocity and head behind it return to their steady-state values. At $t = 4L/a$, the wave arrives at the valve with conditions identical to the initial steady-state conditions, and the process repeats itself with a period of $4L/a$ with no attenuation.

The cyclic nature of the phenomenon is best shown by plotting $H$ versus $t$ for selected locations in the pipe. This is shown in Fig. 8.3a for a point just upstream from the valve and in Fig. 8.3b for the midpoint of the pipeline. Each $4L/a$ seconds, the square wave repeats itself. For real cases with friction, the square wave rapidly turns into a sinusoidal-shaped wave, attenuating with time. The attenuation is caused by energy loss due to fluid friction, expansion of the pipe, and compression of the liquid.

Consider the case of no friction where the flow is completely stopped by reducing the velocity in 3 equal increments, each at $0.25L/a$ apart. Figure

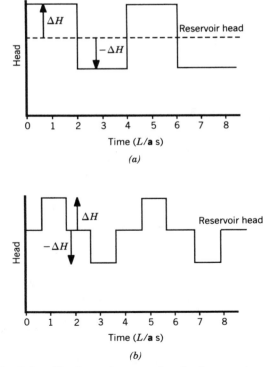

Fig. 8.3a   Head vs. time at valve for instant closure.

Fig. 8.3b   Head vs. time at midpoint for instant closure.

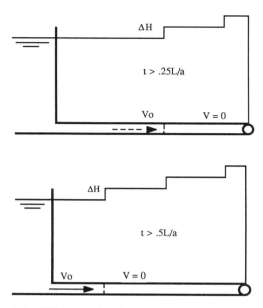

**Fig. 8.4**  HGL at valve for incremental valve closure.

8.4 shows the hydraulic grade lines for the system at $t > 0.25L/a$ and $t > 0.5L/a$. At $t = 0$, the initial head rise is equal to $\Delta H = -a(-0.33V_o)/g$. This wave propagates toward the reservoir. At $t = 0.25$, $L/a$ and $t = 0.5L/a$ additional waves of equal magnitude are generated. The velocity is zero behind the third wave and the total head rise is the sum of the three incremental values or $aV_o/g$.

## 8.4  INSTANT AND EFFECTIVE VALVE CLOSURE TIME

In the previous section, the term instant closure was used to describe the closure time for a valve. Instant closure normally refers to closure over an infinitely small time increment. However, instant closure as it regards closure of a valve at the end of the pipeline, actually refers to a finite time. It is the longest time that a valve can be closed and still cause a pressure rise equal to that of an instant closure.

To explain this, look again at Fig. 8.4. The pressure at the valve is a summation of the head rise caused by the three incremental closures and is equal to $aV_o/g$. Thus, the head rise at the valve is the same magnitude as if the valve were closed instantly. Therefore, as long as the valve is closed in a time less than or equal to $2L/a$ (which is the time required for the first pressure wave to travel to and from the reservoir), the head rise at the valve will be the same as if the valve were closed instantly. The $2L/a$ time is therefore often the instant closure time. For systems with friction, the headrise in $2L/a$ will be less than instant closure.

For a valve at the end of a long pipeline, the instant closure time can be considerably greater than $2L/a$. This is because the friction loss coefficient $fL/d$ is much greater than the loss coefficient for the valve $K_l$ at large openings. Before the valve can significantly reduce the flow, it must close enough that its loss coefficient is a significant percentage of $fL/d$. The effective closure time for a valve is the actual time that a valve is reducing the flow. This time can be much less than the total closure time for a long pipeline. These principles are demonstrated in Example 8.2.

***Example 8.2.*** Consider the pipelines analyzed in Example 4.1. Case A was for a short pipe ($fL/d = 3$) and case B was for a long pipe ($fL/d = 250$). To continue the analysis, assume $d = 10$ in., $a = 1500$ fps, $L_1 = 200$ft., $f_1 = 0.0125$, $L_2 = 17{,}000$ ft., and $f_2 = 0.0123$ for the two cases. Steady-state analysis gives initial velocities of 23.5 and 7.17 fps. The reduction of $Q$ with valve opening for these cases is shown in Fig. 8.5 for a valve closing at constant speed.

For *Case A*, Fig. 8.5 shows that the valve controls the flow over 100% of its valve movement because the flow decreases as the valve first starts to close. Therefore, the effective closure time for the valve is equal to the total closure time. The instant closure time is equal to $2L/a = 400/1{,}500 = 0.27$ s. This is the maximum closure time that gives a pressure rise equal to that of an instant closure.

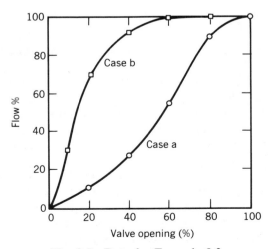

**Fig. 8.5** Data for Example 8.2.

For *Case B*, Fig. 8.5 shows that the flow rate is not significantly reduced until the valve is closed to about 50%. Therefore, the effective closure time is about half the total closure time. For example, if the valve were closed in 200 s, almost no flow reduction would occur during the first 100 s. With $2L/a = 34,000/1,500 = 22.7$ s, the effective instant valve closure time is about 45.4 s since the valve only reduces the flow during the last half of its movement. The longer the pipeline, or larger the $fL/d$, the more a valve must be closed before it begins to reduce the flow.

Next, try to estimate a safe closure time $(t_c)$ for the valve (Case B). To do this, it is necessary to establish a maximum allowable pressure for the pipe. This pressure is dependent on the type and pressure class of pipe, as discussed in Chapter 2. The initial velocity is 7.17 fps and the shutoff pressure is 200 ft or 86.6 psi (refer to Example 4.1). Assume that the maximum allowable pressure is 125 psi. The following simple procedure is an approximate method of estimating valve closure time. A more exact method is presented in Chapter 9 (Example 9.3).

The maximum transient pressure head allowable when the valve is closed the last few degrees is $(125 - 86.6) \cdot 2.31 = 88.7$ ft. The corresponding velocity change $\Delta V = -g\Delta H/a = -32.2 \cdot 88.7/1,500 = -1.9$ fps. This means that if the velocity is reduced 1.9 fps in $2L/a$ or 22.7 s, the transient head rise will be 88.7 ft. From Fig. 8.5, assuming steady-state conditions, find the valve opening for which $V = 1.9$ fps. This represents 26.5% of the original velocity of 7.17 fps and the corresponding valve opening is about 8%. Closing the valve this last 8% in 22.7 s will cause a head rise of 88.7 ft.

If the valve closes at a constant rate, the entire valve closure time would be $22.7/0.08 = 284$ s. For long pipelines, this procedure gives a reasonable estimate. For short pipelines, this calculated time must be increased to allow the transient generated during the previous incremental closure to dampen out.

The closure time could be significantly reduced for this case by rapidly closing the valve to about 30% and then closing at a slower rate.

Note: This problem is solved with a computer program in Example 9.3. The actual time required to keep the maximum head in the pipe to 288 ft is about 190 s.

## 8.5   WAVE PROPAGATION WITH FRICTION

Since all systems contain friction, the pressure wave generated by closure of
a valve will be attenuated. Energy losses associated with friction, expanding
the pipe, compressing the fluid, and any trapped air cause attenuation. To
understand the attenuation process, refer to Fig. 8.6. This figure shows a
pipeline in which essentially all of the reservoir head is dissipated by
friction. The process is easiest to explain by assuming that the transient head
rise is small compared with the reservoir head. At $t = 0$, the valve is instantly
closed, which generates a head rise $\Delta H_1 = aV_o/g$. This pressure wave travels
toward the reservoir at the acoustic wave speed.

Relying on knowledge of the frictionless case, one would assume that the
wave will move up the pipe as shown by the solid curve (a) in Fig. 8.6a. This
produces a sloping HGL parallel to the original HGL. Since the velocity
behind the wave front should be zero, the HGL should be horizontal, as
shown by the broken line (b) in Fig. 8.6a. For this to occur, the pressure at
the valve must increase, which can only occur by packing more water into
the pipe downstream of the wave front (refered to as *line packing*). This
means that there must be some flow in the pipe so that the velocity is not
reduced to zero at the wave front. This makes $\Delta H_2 < \Delta H_1$ and results in a
sloping HGL, as shown in Fig. 8.6b for $t = t_1$.

At the valve, the head continues to rise due to recovery of the friction
loss and the line packing. As the wave continues to move toward the
reservoir, the pressure rise at the wave front continues to reduce. This
process is referred to as *attenuation*. The maximum head at the valve occurs
after $t = L/a$ and is less than reservoir head $\Delta H_1$.

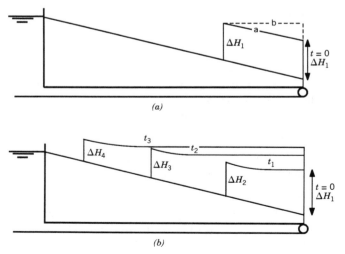

**Fig. 8.6**   Wave propagation with friction. (*a*) incorrect; (*b*) correct.

## 8.6   WAVE SPEED

The transient head rise is directly proportional to the wave speed (Eq. 8.2). Wave speed is therefore a parameter that must be accurately evaluated for each system. Its magnitude is dependent on the density and bulk modulus of the liquid, elasticity, diameter, and wall thickness of the pipe, and the presence of free air or gas.

To understand the physical nature of this dependence, refer to the discussion on propagation of pressure waves (Fig. 8.2). While the initial wave is traveling to the reservoir, liquid still enters the pipe ahead of the wave. The speed with which the wave can move between two arbitrary points in the pipe depends on how much additional liquid can be stored in that short section. The wave travels between the two points in the time it takes for the excess liquid moving at velocity $V_o$, to enter that section and come to rest. If the pipe is very flexible or if there are gas bubbles in the liquid, which makes it more compressible, it will take more liquid and consequently more time to stretch the pipe, compress the liquid or air, and increase the pressure in the pipe. For such a case, the wave travels at a relatively slow speed.

Consider the other extreme of a totally rigid pipe and incompressible liquid. For such a system, the entire column must immediately come to rest when the valve is instantly closed. This hypothetical case would correspond to an infinite wave speed. No fluid is really incompressible and no pipe is totally rigid, so an infinite wave speed is not possible.

The wave speed equation is derived by 1) applying the continuity equation, 2) using Eq. 8.2 to relate $\Delta V$ and $\Delta H$, 3) relating $\Delta H$ to the bulk modulus of the liquid, and 4) relating the expansion of the pipe to its stress–strain properties.

Consider the simple pipe shown in Fig. 8.1$a$. At $t = L/a$ after instant closure of the valve, the velocity in the pipe is zero, the pipe has expanded in diameter and possibly in length, and the fluid is compressed. The total mass entering the pipe is $\rho A V_o L/a$. The mass stored due to elongation of the pipe is small compared with other factors and will be ignored. The amount stored due to increased pipe diameter is $\rho L \Delta A$ and the mass stored due to increased density is $LA\Delta\rho$.

Substituting these into the equation of mass conservation,

$$\frac{\rho A V_o L}{a} = \rho L \Delta A + LA\Delta\rho \tag{8.4}$$

Dividing by $\rho AL$ and replacing $V_o$ (which is equal to $-\Delta V$ for instant closure) with $g\Delta H/a$ from Eq. 8.2.

$$\frac{g\Delta H}{a^2} = \frac{\Delta A}{A} + \frac{\Delta\rho}{\rho}$$

Solving for $a^2$

$$a^2 = \frac{g\Delta H}{\Delta A/A + \Delta\rho/\rho} \tag{8.5}$$

It is desirable to express the wave speed as a function of fluid and pipe properties which are readily obtainable. Using the definition of bulk modulus,

$$K = \frac{\Delta P}{\Delta\rho/\rho} \tag{8.6}$$

and

$$\Delta P = \rho g \Delta H$$

Substituting $g\Delta H = \Delta P/\rho$, multiplying the top and bottom of Eq. 8.5 by $K/\Delta\rho$ and rearranging,

$$a^2 = \frac{K/\rho}{1 + K\Delta A/A\Delta P} \tag{8.7}$$

Next, it is necessary to relate $\Delta A/(A\Delta P)$ to stress–strain properties of the pipe. To be thorough in the derivation one must be concerned about the pipe support conditions. Three conditions are possible:

*Case 1.* The pipe is anchored at the upstream end only.

*Case 2.* The pipe is anchored against any axial movement.

*Case 3.* Each pipe section anchored with expansion joints at each section.

Several sources are available (81, 89) where the complete derivation is provided for each of the cases. The final result is that the pipe constraint usually has less than a 10% influence on the magnitude of the wave speed. It is seldom possible to predict the wave speed with any better accuracy because of uncertainties in both the fluid and pipe properties. It is therefore felt that a simplification of the derivation which ignores any axial strain in the pipe is adequate. The following derivation is specifically for Case 3.

With this limitation, the change in $\Delta A$ will only be a function of circumferential strain. The tensile stress and circumferential strain are related by Young's modulus $E$:

$$\text{Strain} = \text{stress}/E \tag{8.8}$$

The tensile stress for a thin wall pipe is related to the diameter and internal pipe pressure. Figure 8.7 shows a unit long section of pipe of wall thickness $e$ subject to an increased pressure $\Delta P$. The increase in circumferential tensile

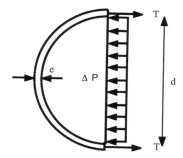

**Fig. 8.7**  Free body of pipe section.

force $T$, stress, and $\Delta P$ are related by conditions of static equilibrium $\Sigma F_x = 0$.

$$2T - d\Delta P = 0 \text{ and } T = e \cdot \text{stress}$$

so

$$\text{Stress} = \frac{d\Delta P}{2e} \tag{8.9}$$

The radial strain is (combining Eqs. 8.8 and 8.9)

$$\text{Strain} = d\Delta P/2eE$$

$\Delta A$ is equal to the circumference of the pipe times the increase in the radius:

$$\Delta A = \pi D \frac{d\Delta P}{2eE} \frac{d}{2} = \frac{\pi d^2}{4} \times \frac{\Delta Pd}{eE} = \frac{A\Delta Pd}{eE} \tag{8.10}$$

Substituting Eq. 8.10 into 8.7 and taking the square root gives

$$a = \frac{\sqrt{K/\rho}}{\sqrt{1 + Kd/Ee}} \tag{8.11}$$

If one wished to include the effect of pipe constraint (81), the equation becomes

$$a = \frac{\sqrt{K/\rho}}{\sqrt{1 + CKd/Ee}} \tag{8.12}$$

$C = 1 - 0.5\mu$ for Case 1 constraint

$C = 1 - \mu^2$ for Case 2 constraint

$C = 1$ for Case 3 constraint

$\mu = $ Poisson's ratio

Equations 8.11 and 8.12 are for circular thin-walled conduits. For noncircular pipes, thick-walled pipes, and other special cases, see References 81 and 87. Values of $K$ and $\mu$ from Reference 81 are listed in Table 8.1.

**Example 8.3.** Calculate the wave speed for an unreinforced concrete pipe carrying water at 20°C. The properties of the pipe are $d = 610$ mm, $e = 76$ mm, $E = 27.6$ GPa, and $\mu = 0.3$ (Table 8.1). For the water, $\rho = 998.2$ kg/m$^3$ and $K = 2.2$ GPa (Table 1.1). Find $a$ for all three pipe constraints.

If the pipe were completely rigid ($E$ = infinity), the wave speed would be

$$a = \sqrt{\frac{K}{\rho}} = \sqrt{\frac{2.2 \times 10^9}{998.2}} = 1483 \text{ m/s}$$

For the actual case, $Kd/Ee = 0.64$.

*Case 1.* Pipe anchored at upstream end only:

$$C = 0.85$$
$$a = \frac{1483}{\sqrt{1 + 0.64 \times 0.85}} = 1193 \text{ m/s}$$

*Case 2.* Pipe anchored against any axial movement:

$$C = 0.91$$
$$a = \frac{1483}{\sqrt{1 + 0.64 \times 0.91}} = 1179 \text{ m/s}$$

**TABLE 8.1   Moduli of Elasticity and Poisson's Ratio for Common Pipe Materials[a]**

| | | |
|---|---|---|
| Steel | $E = 30 \times 10^6$ psi | $\mu \approx 0.30$ |
| Ductile cast iron | $E = 24 \times 10^6$ psi | $\mu \approx 0.28$ |
| Copper | $E = 16 \times 10^6$ psi | $\mu \approx 0.30$ |
| Brass | $E = 15 \times 10^6$ psi | $\mu \approx 0.34$ |
| Aluminum | $E = 10.5 \times 10^6$ psi | $\mu \approx 0.33$ |
| PVC | $E \approx 4 \times 10^5$ psi | $\mu \approx 0.45$ |
| Fiberglass reinforced | $E_2 = 4.0 \times 10^6$ psi | $\mu_2 = 0.27 \cdot 0.30$ |
| plastic (FRP) | $E_1 = 1.3 \times 10^6$ psi | $\mu_1 = 0.20 \cdot 0.24$ |
| Asbestos cement | $E \approx 3.4 \times 10^6$ psi | $\mu \approx 0.30$ |
| Concrete | $E = 57{,}000\sqrt{f_c'}$[b] | $\mu \approx 0.30$ |

[a]Reference 81. (Reproduced from Gary Z. Walton, "*Analysis and Control of Unsteady Flow in Pipelines*," Second Ed. (Stoneham, MA: Butterworth Publishers, 1984). With permission from the publishers.)
[b]$f_c'$ = 28-day strength in psi.

*Case 3.* Expansion joints:

$$C = 1.0$$

$$a = \frac{1483}{\sqrt{1 + 0.64}} = 1158 \text{ m/s}$$

The example shows how little the type of constraint influences the wave speed.

***Example 8.4.*** Compare the head rise caused by instant closure of a valve if the initial velocity is 5.24 fps for: (a) a steel pipeline $E = 29 \times 10^6$ psi, $e = 0.5$ in., $d = 12$ in.; and (b) a plastic pipeline $E = 4 \times 10^5$ psi, $e = 0.5$ in., $d = 12$ in. Use $K = 3.2 \times 10^5$ psi and $\rho = 1.94$ slug/ft$^3$.

(a) steel pipe: $Kd/Ee = 0.264$, from

$$a = \frac{\sqrt{144 \times 3.2 \times 10^5/1.94}}{\sqrt{1 + 0.264}} = 4335 \text{ fps}$$

so

$$\Delta H = 4335 \times 5.74/32.2 = 773 \text{ ft}$$

(b) plastic pipe: $Kd/Ee = 19.2$, from

$$a = \frac{4874}{\sqrt{1 + 19.2}} = 1084 \text{ fps}$$

so

$$\Delta H = \frac{1084 \times 5.74}{32.2} = 193 \text{ ft}$$

The steel pipe has almost four times the head rise because of the higher wave speed. The influence of wave speed on head rise decreases as the valve closure time increases. This is demonstrated in Example 9.6.

## 8.7  INFLUENCE OF AIR ON WAVE SPEED

Example 8.4 demonstrates one of the methods of controlling transients for cases in which the valve closure is almost instantaneous, reducing the wave speed by using a more flexible pipe. Another method of reducing the wave speed is to reduce the effective bulk modulus of the water so it is more compressible. This is done by introducing a small amount of air into the system or by installing an air chamber. Equation 8.6 can be changed to $K = -\Delta P/\Delta v \,/v$. If there is a small amount of undissolved air in a volume of water $v$, a pressure increase of $\Delta P$ will cause a larger $\Delta v$ because the air easily compresses. The result is a large decrease in $K$ and consequently a decrease in the wave speed.

An equation for predicting the wave speed of an air–water mixture was developed by Streeter (75). He applied continuity and momentum to a small control volume. After eliminating the smallest term, the resulting equation is

$$a = \sqrt{\frac{K/\rho}{1 + Kd/Ee + MR_gKT/P^2}} \qquad (8.13)$$

in which $M$ = mass of air per unit volume of mixture; $P$ = absolute pressure; $T$ = absolute temperature; $R_g$ = gas constant $K$ = bulk modulus.

***Example 8.5.*** A 12-in. diameter 0.25-in. wall steel pipe is filled with aerated water ($M = 10^{-5}$ slugs/ft$^3$). Calculate the wave speed at an atmospheric pressure of $P_b = 14$ psia. $T = 60°F$.

$$K = 3.11 \times 10^5 \text{psi (Table 1.1)}$$

$$\rho = 1.94 \text{ slugs/ft}^3 \text{ (Table 1.1)}$$

$$R_g = 1715 \text{ ft-lb/slug°Rankine}$$

$$P = 14 \cdot 144 = 2016 \text{ lb/ft}^2 \text{ (absolute)}$$

$$E = 30 \times 10^6 \text{psi (Table 8.1)}$$

$$Kd/Ee = 0.498$$

$$MR_gT \ K/P^2 = 98.3$$

From Eq. 8.13

$$a = 480 \text{ fps}$$

To demonstrate the dependence of $a$ on the mass of air and the absolute pressure, Eq. 8.13 has been plotted in Fig. 8.8. The reduction in $a$ due to increased air is observed by looking at the intersections of the curves for different amounts of air with a vertical line of constant head. For example, from Fig. 8.8 at $H_{abs} = 8$ m, the wave speed drops from about 1260 m/s with essentially no air to about 80 m/s with 1% air (calculated at standard conditions).

The influence of pressure is observed by following a line of constant air mass. For 0.1% air, $a$ varies from about 30 m/s to over 1200 m/s as the absolute pressure head is increased from 1 to 128 m. It turns out that the presence of the air not only reduces the wave speed, but causes significant energy dissipation. The principle of reduced wave speed is simple, but applying it to the numerical solution of real transient problems is complex. There is also a practical problem associated with applying this principle, since knowing and controlling

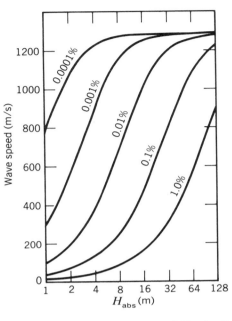

**Fig. 8.8** Wave speed for air–water mixtures (air at standard temperature and pressure; $KD/Ee = 0.263$).

the amount of air is difficult. Problems related to air in pipelines and computational schemes currently being used to calculate transients with free air will be discussed in Chapter 10.

## PROBLEMS

**8.1.** Why does closing a valve slower reduce the transient pressure rise?

**8.2.** Referring to Fig. 8.2, draw the HGL at $t = 3.5L/a$.

**8.3.** Referring to Fig. 8.3, draw $H$ versus $t$ at $x = 0.25L$.

**8.4.** Discuss the significance of the times $2L/a$ and $4L/a$.

**8.5.** Define instant valve closure and effective valve closure.

**8.6.** For a long pipe with friction (such as Fig. 8.6), plot the HGL for $t = 1.5L/a$ and $2.5L/a$ (assume instant closure).

**8.7.** Draw the $H$ versus $t$ curve at the valve for a pipe with high friction for time up to $2L/a$ (assume instant closure).

**8.8.** Find the wave speed for an 8-in. PVC pipe, $e = 0.5$-in. (the water is at 80°F and the pipe has expansion joints).

**8.9.** Outline the steps in the derivation of the wave speed equation.

**8.10.** Define attenuation and line packing.

**8.11.** For instant valve closure at the end of a long pipeline with friction (Fig. 8.6), will the maximum head in the pipe be $HR + \Delta H_1$? ($HR =$ reservoir head). If $\Delta H_5 =$ the head rise of the wave front at $t = L/a$, how will the maximum head at the valve compare to $HR + \Delta H_5$?

# 9

# NUMERICAL SOLUTIONS
# OF TRANSIENTS

## 9.1 METHOD OF ANALYSIS

Computational techniques were introduced in the previous chapter for estimating pressures for simple transients. Most practical situations involve piping systems and changing flow conditions that are too complex to be handled with simple hand calculations. In this chapter, the equations that govern unsteady flow in pipelines are developed. The resulting partial differential equations are solved by the method of characteristics, which transforms them into total differential equations. After integration, the equations are solved numerically by finite differences. This analysis provides equations that can be used to predict the flow and head at any interior pipe section at any time. To complete the analysis, equations describing all boundary conditions are required. Typical boundary conditions are connection of a pipe to a reservoir, a valve, changes in pipe diameter or material, pipe junctions, etc. Friction loss is included in the development of the basic equations and minor losses are handled as boundary conditions.

The analysis is developed using elastic theory where compressibility of the fluid and elasticity of the pipe are included. It properly models the propagation and reflections of the pressure wave. It can also be used for surge calculations although, computing times would be longer due to the small time step inherent with transient calculations. With numerical surge calculations, the time interval can have an order of magnitude larger than is required for transients.

For some problems, transient analysis is used for part of the system and surge analysis for other parts. One example is a hydropower installation where there is a power tunnel, surge tank, and penstock. When rapid changes in flow are initiated at the turbine (as is the case of load rejection), a transient is generated in the penstock. The surge tank acts as a reservoir and prevents most of the transient pressures from entering the power tunnel. The flow in the power tunnel is handled by rigid column or surge analysis and the penstock with elastic or transient analysis.

## 9.2   EQUATION OF MOTION

Unsteady flow is governed by the equation of motion, which is that the summation of all forces acting on a mass of fluid in a given direction is equal to the product of the mass and the acceleration in the direction of the force. Figure 9.1 is a free-body diagram showing forces acting on a cylindrical segment of fluid. The forces include pressure forces acting on the two ends, friction or shear forces on the outer surface, and gravity. The equation of motion applied in the direction of the fluid motion $x$ is

$$PA - \left(P + \frac{\partial P}{\partial x}\,\Delta x\right) A - \tau\pi D\Delta x + \rho g A\Delta x \sin\theta = \rho A\Delta x \frac{DV}{Dt}$$

Simplifying and dividing by $\Delta x$ gives:

$$-\frac{\partial P}{\partial x}\,A - \tau\pi D + \rho g A \sin\theta = \rho A \frac{DV}{Dt} \qquad (9.1)$$

The wall shear stress for unsteady flow is normally assumed to be equal to the steady-state value at the same velocity. It is customary to work with a friction loss expressed as a loss of head over a reach of pipe rather than dealing with a distributed wall shear stress. This relationship can be developed by applying the equation of motion to Fig. 9.1 for steady-state flow for the simplified case of $\theta = 0$.

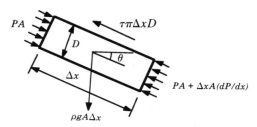

**Fig. 9.1**   Free-body diagram of fluid section.

$$(P_1 - P_2)A = \tau \pi D \Delta x = \gamma(H_1 - H_2)A \qquad (9.2)$$

From the Darcy–Weisbach equation:

$$H_1 - H_2 = \frac{fLV^2}{2gD} = \frac{f\Delta x \, V|V|}{2gD} \qquad (9.3)$$

in which $f$ is the friction factor. The absolute value sign on $V$ is to ensure that the friction force is always opposite to the direction of flow. Substituting Eq. 9.3 into 9.2 and rearranging the shear stress term in Eq. 9.1,

$$\tau \pi D = \frac{\rho f V |V| A}{2D} \qquad (9.4)$$

Dividing Eq. 9.1 by $\rho A$, substituting in Eq. 9.4 and expanding the total derivative of $V$,

$$\frac{\partial P}{\rho \partial x} + \frac{fV|V|}{2D} - g \sin \theta + \frac{\partial V}{\partial x}\frac{dx}{dt} + \frac{\partial V}{\partial t} = 0 \qquad (9.5)$$

This can be further simplifed by using the piezometric head $H = P/\gamma + z$ (see Fig. 9.1). Differentiating gives

$$\frac{\partial H}{\partial x} = \frac{\partial P}{\gamma \partial x} + \frac{dz}{dx} \qquad (9.6)$$

From Fig. 9.1 it is seen that $dz/dx = -\sin \theta$. Consequently, Eq. 9.5 can be reduced to

$$g\frac{\partial H}{\partial x} + \frac{fV|V|}{2D} + \frac{V\partial V}{\partial x} + \frac{\partial V}{\partial t} = 0 \qquad (9.7)$$

The use of the piezometric head in Eq. 9.7 restricts its application to liquids. It is important to remember that all further computations must be made using the piezometric head, not the pressure head. For a more general derivation, see Section 2.1 of Reference 87.


## 9.3  EQUATION OF CONTINUITY

The law of conservation of mass applied to a control volume states that the net mass flux through the control surface must equal the time rate of change of mass inside the control volume. When applying this to the present situation, it is noted that the control volume (in this case a short section of pipe) can increase in cross-sectional area and length due to increased pressure. For the development that follows, it is assumed that the lengthening of the pipe is negligible. The continuity equation becomes

$$\rho AV - \left[ \rho AV + \frac{\partial(\rho AV)}{\partial x} \, dx \right] = \frac{\partial(\rho A dx)}{\partial t}$$

or

$$-\frac{\partial(\rho AV)}{\partial x} \, dx = \frac{\partial(\rho A dx)}{\partial t} \tag{9.8}$$

Expanding both sides of Eq. 9.8,

$$-\left( \rho A \frac{\partial V}{\partial x} \, dx + \rho V \frac{\partial A}{\partial x} \, dx + AV \frac{\partial \rho}{\partial x} \, dx \right) = \rho dx \frac{\partial A}{\partial t} + A dx \frac{\partial \rho}{\partial t}$$

Rearranging and dividing by $\rho A dx$,

$$\frac{1}{A} \left( \frac{\partial A}{\partial t} + V \frac{\partial A}{\partial x} \right) + \frac{1}{\rho} \left( \frac{\partial \rho}{\partial t} + V \frac{\partial \rho}{\partial x} \right) + \frac{\partial V}{\partial x} = 0$$

The first two terms in parentheses are the total derivatives of $A$ and $\rho$ with respect to time, so the equation reduces to

$$\frac{1}{A} \frac{dA}{dt} + \frac{1}{\rho} \frac{d\rho}{dt} + \frac{\partial V}{\partial x} = 0 \tag{9.9}$$

Next, replace $dA/dt$ with structural properties of the pipe as was done for the development of the wave speed equation in Section 8.6. Expressing Eq. 8.10 in differential form and rearranging,

$$dA = \frac{AD}{eE} \, dP = \frac{\rho g A D dH}{eE}$$

The first term of Eq. 9.9 then becomes

$$\frac{\rho g D}{eE} \frac{dH}{dt}$$

The second term in Eq. 9.9 can be expressed in terms of the bulk modulus and $dH$ using the definition

$$K = \frac{dP}{d\rho/\rho} \text{ or } \frac{d\rho}{\rho} = \rho g \frac{dH}{K}$$

The second term in Eq. 9.9 becomes

$$\frac{\rho g}{K} \frac{dH}{dt}$$

Substituting into Eq. 9.9 and rearranging,

$$\frac{dH}{dt} \left( \frac{1 + KD/eE}{K/\rho} \right) + \frac{1}{g} \frac{\partial V}{\partial x} = 0 \qquad (9.10)$$

Using Eq. 8.10 for wave speed, Eq. 9.10 reduces to

$$\frac{dH}{dt} + \frac{a^2}{g} \frac{\partial V}{\partial x} = 0 \qquad (9.11)$$

This is the final form of the continuity equation which will be solved together with the equation of motion (Eq. 9.7) since they provide two equations in two unknowns $H$ and $V$. The technique to be used to transform the partial differential equations into total differentials is the method of characteristics.

## 9.4 METHOD OF CHARACTERISTICS

The momentum and continuity equations just developed are (expanding $dH/dt$ in Eq. 9.11)

$$g \frac{\partial H}{\partial x} + \frac{fV|V|}{2D} + \frac{V\partial V}{\partial x} + \frac{\partial V}{\partial t} = 0 \qquad (9.7)$$

$$V \frac{\partial H}{\partial x} + \frac{\partial H}{\partial t} + \frac{a^2}{g} \frac{\partial V}{\partial x} = 0 \qquad (9.12)$$

These can be simplified by comparing the relative magnitudes of the various terms and eliminating those of lesser importance. Looking at the equations in finite difference form and recognizing that

$$\frac{\Delta V}{\Delta t} \text{ and } g \frac{\Delta H}{\Delta x} > V \frac{\Delta V}{\Delta x}$$

$$\frac{\Delta H}{\Delta t} \text{ and } a^2 \frac{\Delta V}{\Delta x} \gg V \frac{\Delta H}{\Delta x}$$

the equations simplify to

$$g \frac{\partial H}{\partial x} + \frac{\partial V}{\partial t} + \frac{fV|V|}{2D} = 0 \qquad (9.13)$$

$$\frac{\partial H}{\partial t} + \frac{a^2}{g} \frac{\partial V}{\partial x} = 0 \qquad (9.14)$$

By multiplying Eq. 9.14 by an unknown constant $\lambda$, adding it to Eq. 9.13, and rearranging,

$$\lambda \left( \frac{g}{\lambda} \frac{\partial H}{\partial x} + \frac{\partial H}{\partial t} \right) + \left( \frac{\lambda a^2}{g} \frac{\partial V}{\partial x} + \frac{\partial V}{\partial t} \right) + \frac{fV|V|}{2D} = 0 \qquad (9.15)$$

Since Eq. 9.15 is a linear combination of two independent equations, any two real values of $\lambda$ will produce two equally independent equations. The values are chosen to make the equations total derivatives. From the definition of a total derivative,

$$\frac{dH}{dt} = \frac{\partial H}{\partial x} \frac{dx}{dt} + \frac{\partial H}{\partial t} \quad \text{and} \quad \frac{dV}{dt} = \frac{\partial V}{\partial x} \frac{dx}{dt} + \frac{\partial V}{\partial t} \qquad (9.16)$$

Comparison of Eq. 9.16 with 9.15 reveals the values of $\lambda$.

$$\frac{dx}{dt} = \frac{g}{\lambda} = \frac{\lambda a^2}{g} \quad \text{or} \quad \lambda = \pm \frac{g}{a} \qquad (9.17)$$

so

$$\frac{dx}{dt} = \pm a$$

With these two $\lambda$ values, Eq. 9.15 becomes two independent total differential equations:

$$\frac{g}{a} \frac{dH}{dt} + \frac{dV}{dt} + \frac{fV|V|}{2D} = 0 \qquad C^+ \text{equation} \qquad (9.18)$$

$$\text{for } \frac{dx}{dt} = +a \qquad (9.19)$$

$$\frac{g}{a} \frac{dH}{dt} - \frac{dV}{dt} - \frac{fV|V|}{2D} = 0 \qquad C^- \text{equation}$$
$$(9.20)$$

$$\text{for } \frac{dx}{dt} = -a \qquad (9.21)$$

Equations 9.18 and 9.20 are called the $C^+$ and $C^-$ compatibility equations (87).

## 9.5   FINITE DIFFERENCE SOLUTION

Solution of Eqs. 9.18 and 9.20 is done using finite differences. Figure 9.2a shows a simple reservoir valve system. Divide the pipe into $N$ equal sections of length $\Delta x$. For this example $N = 4$, so there are $N + 1 = 5$ nodes where calculations will be made.

Before any transient is generated, the head $H_I$ and flow $Q_I$ at the five nodes can be found by applying the energy equation to the system. For the frictionless case where all the reservoir head is dissipated by the downstream

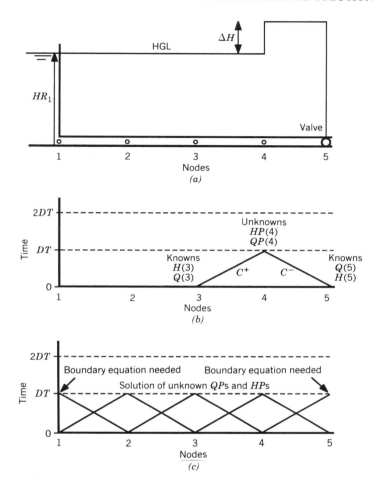

**Fig. 9.2**  Graphical solution of the $C^+$ and $C^-$ equations.

valve, $Q_I = QI$ and $H_I = HR_1$ at all five nodes, where $QI$ is the initial flow and $HR_1$ is the piezometric head at the reservoir.

A transient is generated at time $t$ by instant closure of the valve at node 5. This creates a wave of magnitude $H = aVo/g$ (Eq. 8.1) which travels toward the reservoir at the speed of sound $a$. It arrives at node 4 at time $t + \Delta t = \Delta t = \Delta x/a$, $\Delta x = L/a$ (the distance between nodes). The $C^-$ equation (9.20) transmits information about changes in $H$ and $Q$ to upstream nodes since it is valid for $dx/dt = -a$ (Eq. 9.21). Similarly, the $C^+$ equation (9.18) transmits information in the $+a$ direction.

Solution of the transient equations consists of finding values of head and flow at each node as the transient progresses. Calculations will be made at each $\Delta t$ time interval. The nodes are equally spaced at a distance $\Delta x$ and the wave speed is constant. This means that the time $\Delta t$ for the pressure wave to travel between any two nodes is the same.

The unknown heads and discharge at the nodes at time $t + \Delta t$ are labeled $HP_I$ and $QP_I$. The known head and discharge at the previous time step are $H_I$ and $Q_I$. Consider the transient shown in Fig. 9.2a where the wave has just arrived at node 4. Figure 9.2b is a graphical representation of the solution of Eqs. 9.18–9.21 for node $N$ ($I = 4$). The unknown head $HP_I$ and flow $QP_I$ at node $I$ at time $t + \Delta t$ can be calculated by integrating Eqs. 9.18 and 9.20.

Before integrating, multiply both equations by $adt/g$, change from $V$ to $Q$, and replace $dt$ by $dx$ ($dx = adt$). For the $C^+$ equation, integrating from node $(i - 1)$ to node $(i)$ gives

$$\int_{H_{I-1}}^{HP_i} dH + \frac{a}{gA} \int_{Q_{I-1}}^{QP_I} dQ + \frac{f}{2gDA^2} \int_{0}^{\Delta x} Q|Q|dx = 0 \qquad (9.22)$$

To integrate the third term, the variation of $Q$ with $x$ must be specified. For many problems, using a first-order assumption ($Q = Q_{I-1}$) is adequate. With this assumption, integration of Eq. 9.22, and a similar treatment of the $C^+$ equation.

$$C^+: \quad HP_I - H_{I-1} + B[QP_I - Q_{I-1}] + R\ Q_{I-1}|Q_{I-1}| = 0 \qquad (9.23a)$$

$$C^-: \quad HP_I - H_{I+1} - B[QP_I - Q_{I+1}] - R\ Q_{I+1}|Q_{I+1}| = 0 \qquad (9.23b)$$

in which

$$B = \frac{a}{gA} \qquad (9.24)$$

and

$$R = \frac{f\Delta x}{(2gDA^2)} \qquad (9.25)$$

Since $H_{I-1}$, $H_{I+1}$, $Q_{I-1}$ and $Q_{I+1}$ are known and $B$ and $R$ are constants, the equations can be further simplified to

$$C^+: \quad HP_I = CP - B\ QP_I \qquad (9.26)$$

$$C^-: \quad HP_I = CM + B\ QP_I \qquad (9.27)$$

$$CP = H_{I-1} + B\ Q_{I-1} - R\ Q_{I-1}|Q_{I-1}| \qquad (9.28)$$

$$CM = H_{I+1} - B\ Q_{I+1} + R\ Q_{I+1}|Q_{I+1}| \qquad (9.29)$$

Adding Eqs. 9.26 and 9.27 gives

$$HP_I = \frac{(CP + CM)}{2} \tag{9.30}$$

$QP_I$ can be found by substituting $HP_I$ back into Eq. 9.26 or 9.27. Refer again to Fig. 9.2b. The $C^+$ equation (9.26) provides information from node $I - 1$, or node 3 in this case, to node 4 along the $C^+$ line. Similarly, Eq. 9.27 provides information along the $C^-$ line. Solving Eq. 9.30 at time $t + \Delta t$ provides $HP_4$.

Since $HP$ and $QP$ are desired at each node the process is repeated at each node. Figure 9.2c shows this graphically. All interior nodes $(I = 2, N)$ transmit both $C^+$ and $C^-$ information to adjacent nodes. With additional boundary conditions for nodes 1 and $N + 1$, $HP$ and $QP$ at each node can be found at $t + \Delta t$. The next step is to set $Q = QP$ and $H = HP$, recalculate $CP$ and $CM$, and solve $HP$ and $QP$ again at each succeeding time step.

To aid in understanding the method of solution, consider that an observer is placed at each node. At each time interval $\Delta t$ they are to transmit information to adjacent nodes (or observers) about the head and flow conditions at their node.

Refer again to Fig. 9.2a. At time $t$, the valve is closed and the observer at node 5 notes that the flow went to zero, $Q_5 = 0$, and the head increased by $\Delta H = aV_o/g$, so $H_5 = HR_1 + \Delta H$. At all other nodes, the observers see no change. This information is sent to each adjacent observer. Observer 4 can expect to see the pressure wave generated at node 5 arrive at his location at $t + \Delta t$. Since observer 3 reported no change, the net result at node 4 at $t + \Delta t$ will be $QP_4 = 0$ and $HP_4 = HR_1 + \Delta H$. This is demonstrated in the following example.

**Example 9.1.** For the transient shown in Fig. 9.2a, calculate $HP_4$ and $QP_4$ at $t + \Delta t$. This is done by solving Eqs. 9.26–9.30 for $I = 4$.

First evaluate $CP$ and $CM$ at the instant the valve is closed realizing that $Q_3 = QI$ (initial flow), $H_3 = HR_1$, $Q_5 = 0$, $H_5 = HR_1 + \Delta H$, $\Delta H = B\ QI$ and ignoring friction $(R = 0)$:

$$CP = HR_1 + B\ QI$$
$$CM = HR_1 + \Delta H = HR_1 + B\ QI$$

Substitute these values into Eqs. 9.26 and 9.27:

$$HP_4 = HR_1 + B\ QI - B\ QP_4$$
$$HP_4 = HR_1 + B\ QI + B\ QF_4$$

Solving (or using Eq. 9.30 directly) gives

$$HP_4 = HR_1 + B\ QI = HR_1 + HP_4 = HR_1 + \Delta H$$

Substituting $HP_4$ into either Eq. 9.26 or 9.27 gives $QP_4 = 0$. This is precisely what would be expected from the description of pressure-wave propagation discussed in Section 8.3.

Equations (9.26–9.29) must also hold for steady state flow where $Q_{i-1} = Q_{i+1} = QI$. Equations 9.26 and 9.27 for this case (including friction) reduce to

$$HP_I = H_{I-1} - R\ QI|QI|$$

$$HP_I = H_{I+1} + R\ QI|QI|$$

Since $R\ QI|QI|$ is the friction loss in a reach $\Delta x$, steady state is satisfied.

Thus far, equations (9.26–9.29) have been developed that can calculate the head and flow at any time step for all interior points (nodes 2 to $N$). At nodes 1 and $N + 1$, there are still two unknowns but only one equation from the adjacent node available, so additional information at the boundaries is needed.

## 9.6  SIMPLE BOUNDARY CONDITIONS

### Reservoir

For the system shown in Fig. 9.2a, the boundary condition at node 1 is a reservoir. If it is large so the level does not change during the transient, the head will be constant so $HP_1 = HR_1 =$ piezometric head at reservoir. The flow $QP_1$ is found from the $C^-$ characteristic (Eq. 9.27) with $CM$ calculated at node 2.

### Valve at End of Pipe

For a valve at the end of the pipe (node $= NS = N + 1$), the boundary condition is the equation for head loss across the valve. To make the solution general, assume there is a reservoir of known elevation downstream. Let $HR_2$ be the piezometric head in the reservoir downstream from the valve. Neglect pipe exit loss and, using Eqs. 1.29 or 4.1 for loss across the valve,

$$HP_{NS} = HR_2 + \frac{K_l QP_{NS}|QP_{NS}|}{2gA_v^2} \tag{9.31}$$

with $C3 = K_l/(2gA_v^2)$, $Av = 0.7854\ d^2$, $d =$ valve diameter. (Use $+ C3$ for forward flow and $- C3$ for reverse flow.) The eq. reduces to

$$HP_{NS} = HR_2 + C3\ QP_{NS}^2 \tag{9.32}$$

This can be solved with the $C^+$ equation (9.26) to give

$$HR_2 + C3\ QP_{NS}^2 = CP - QP_{NS}$$

or

$$QP_{NS}^2 + CC3\ QP_{NS} + CC4 = 0$$

in which $CC3 = B/C3$, $CC4 = (HR_2 - CP)/C3$

Solution of the quadratic gives a real root of

$$QP_{NS} = 0.5(-CC3 \pm \sqrt{CC3^2 - 4\ CC4}) \tag{9.33}$$

For forward flow, the choice of $+\sqrt{}$ as the real solution is based on the fact that $B/C3$ is always positive, so to get a positive $QP_{NS}$ requires using the positive sign in the equation. For reverse flow, change the sign of $C3$ and use the negative sign in eq. 9.33. With $QP_{NS}$ known from Eq. 9.33, $HP_{NS}$ is calculated with the $C^+$ equation (9.26).

Solving the valve boundary requires that $K_l$ be known at each time interval. This requires specifying the variation of valve opening with time (usually a linear function) and consequently the variation of $K_l$ with time. Since valve data are often given in terms of $C_d$ (Eq. 4.4) or $C_v$ (Eq. 4.2), one must first convert to $K_l$ with Eqs. 4.6 or 4.11 and 4.6.

For a free-discharge valve, $HR_2 = Z_{NS}$ and $K_l$ must be adjusted to compensate for the lack of pressure recovery downstream. Actually, use $C_{df}$ (Eq. 4.5).

Typical $C_d$ data for valves are shown in Fig. 4.3, in which it can be seen that the shape of the $C_d$ curves for different valves varies significantly. Consequently, they cannot be accurately represented by a simple functional relationship. The method recommended is to select $C_v$ or $C_d$ values (and calculate $K_l$ at each 10% of valve movement (11 values total) and linearly interpolate between them as the valve closes. The valve can be set to start at any valve opening by selecting the first $C_d$ at that point.

To simulate opening a valve, the calculation of the initial conditions will need to be changed if the valve starts fully closed, since all flows are zero and all heads are equal to the reservoir head. The other modification is to enter the $K_l$ data in the reverse order (starting from fully closed).

With the reservoir and valve boundary conditions solved, and using them with the $C^+$ and $C^-$ equations, $HP_I$ and $QP_I$ at all nodes can be calculated at $t + \Delta t$. The next step is to re-index so that these new values become the known conditions to get the $CP$ and $CM$ values of Eqs. 9.28 and 9.29. Then repeat the process to find $HP_I$ and $QP_I$ at all subsequent time steps.

## 9.7  BASIC WATERHAMMER PROGRAM

A basic FORTRAN program for a single pipeline with a reservoir and valve (Fig. 9.2a) is listed in Fig. 9.3. The program consists of six parts: 1) dimensioning the variables, identifying the input data file (tran.dat), reading in the data, and calculating the constants (lines 1–16); 2) calculating initial conditions by equating the difference in reservoir levels to the friction and valve loss (lines 17–19); 3) solving the interior points (lines 25–29); 4) solving the upstream boundary (lines 30–32); 5) solving the valve boundary condition by first interpolating to get the discharge coefficient and then solving Eqs. 9.33 and 9.26 (lines 33–47); and 6) reinitializing the data (lines 48–50). Note that some versions of FORTRAN cannot accept the ! comment. The program is very simple to demonstrate the basic features of the procedure. The valve must fully close.

> **Example 9.2.** Use the program listed in Fig. 9.3 to solve the transient caused by valve closure in the simple pipeline shown in Fig. 9.2a for (a) instant closure and no friction ($f = 0$); (b) instant closure with friction ($f = 0.03$); and (c) closure in $t_c = 2L/a$ s with friction. Use a horizontal pipe with the datum through the pipe centerline $XL = 2000$ ft, $N = 4$, $D = 1$ ft, $f = 0$, $HR_1 = 50$, $HR_2 = 0$, $a = 1500$ fps, $g = 32.2$ fps$^2$, $t_c = 0$, $t_{max} = 5$ s. The 11 $C_d$ values would be selected from data such as Fig. 4.3 for the proper valve and for the desired range of valve openings. Have it initially open 50%, so $C_{d1} = 0.24$ (initial value). The other 10 values are read at equally spaced value openings between 50% and 0.

The values of $C_d$ selected, together with all input data, are stored in data file tran.dat and are printed out at the top of Tables 9.1–9.3. Table 9.1 lists the input data and results of running the program in Fig. 9.3 for instant valve closure for a frictionless pipe (Case a). The first two lines of $Q$ and $H$ of output data at $t = 0$ are for steady state. Once the valve is closed at $t = 0.333$ s, the head at node $NS$ increases by $aV_o/g = 653$ ft. This pressure wave travels to the reservoir, as shown by the diagonal line. Once it hits the reservoir, the wave is reflected back to the valve along the second diagonal line. It is helpful to correlate this output with the description of the wave reflections shown in Figs. 8.2 and 8.3 and with Example 9.1. Once the reflected wave arrives back at the valve ($t = 3$ s) the head drops to $50 - 653$ ft $= -603.52$ ft. Since vapor pressure is the lowest possible pressure, the program is invalid for $t > 3$ s.

When running a program for the first time, some of the following items may help in debugging. First, run the program for steady state by setting $t_c$ very large or having all $C_d$ values the same. The output should show no change with time. Next, run a frictionless case for instant closure, as in Table 9.1. Check that the head rise is $a\Delta V/g$ and that the wave travels as it should through the pipe.

```
1          DIMENSION CD(11),Z(5),HP(5),H(5),QP(5),Q(5)
2          OPEN(5,FILE='TRAN.DAT',STATUS='OLD')
3          READ(5,*,END=83)CD,Z,XL,N,D,f,RH1,RH2,a,g,TC,TMAX
4          NS=N+1                            !number of nodes
5          HR1=RH1+Z(1)                      !piezometric head
6          HR2=RH2+Z(NS)                     !piezometric head
7          t=0                               !initialize time
8          DL=XL/N                           !length of pipe reach
9          Dt=DL/a                           !time step
10         AR=0.7854*D**2                    !pipe area
11         B=a/g/AR                          !transient coef.
12         CC=CD(1)                          !initial flow coef
13         ZK=(1/CC**2)-1                    !valve loss coef.
14         C1=f*XL/(2*g*D*AR**2)             !friction coef.
15         C2=ZK/(AR**2*2*g)                 !valve loss coef.
16         R=C1/N                            !friction coefficient
17         QI=SQRT((HR1-HR2)/(C1+C2))        !initial flow
17a        Do 73 I=1,NS
18         H(i)=HR1-(f*(I-1)*DL/D)*QI**2/(2*g*AR**2)
19 73      Q(i)=QI
20 62      WRITE(9,65)t,(Q(I),I=1,5),(H(I),I=1,5)
21         WRITE(6,65)t,(Q(I),I=1,5),(H(I),I=1,5)
22 65      FORMAT(T2,'t=',F6.2,2X,'Q=',5F10.2/T12,'H=',5F10.2)
23         t=t+Dt
24         IF (t.GT.TMAX)GO TO 83
25         DO 10 I=2,n                       !interior pionts
26         CP=H(I-1)+B*Q(I-1)-R*Q(I-1)*ABS(Q(I-1))
27         CM=H(I+1)-B*Q(I+1)+R*Q(I+1)*ABS(Q(I+1))
28         HP(I)=.5*(CP+CM)
29 10      QP(I)=(CP-HP(i))/B
30         HP(1)=HR1                         !upstream reservoir
31         CM=H(2)-B*Q(2)+R*Q(2)*ABS(Q(2))
32         QP(1)=(HP(1)-CM)/B
33         CP=H(N)+B*Q(N)-R*Q(N)*ABS(Q(N))   !valve at end of pipe
34         IF(t.LT.TC)THEN
35         DTV=TC/10.                        !interpolate to get CC
36         I=IFIX(t/DTV)+1
37         TH=(t-(I-1)*DTV)/DTV
38         CC=CD(I)+TH*(CD(I+1)-CD(I))
39         C3=((1/CC**2)-1)/(2*g*AR**2)
40         CC3=B/C3
41         CC4=(HR2-CP)/C3
42         QP(NS)=.5*(-CC3+SQRT(CC3**2-4.*CC4))
43         HP(NS)=CP-B*QP(NS)
44         ELSE
45         QP(NS)=0
46         HP(NS)=CP
47         END IF
48         DO 11 I=1,NS                      !reinitialize values
49         Q(I)=QP(I)
50 11      H(I)=HP(I)
51         GO TO 62
52 83      STOP
53         END
```

**Fig. 9.3** Waterhammer program.

**TABLE 9.1    Data for Example 9.2, Case a**

*Input Data*

* PIPE LENGTH = 2,000 ft
* NUMBER OF PIPE SECTIONS = 4
* PIPE DIA. = 1.00 ft
* VALVE DIA. = 1.00 ft
* FRICTION FACTOR = 0.000
* WAVE SPEED = 3000 ft
* VALVE OPER. TIME = 0 s
* PROG. RUN TIME = 3.0 s
* US. RES. HEAD = 50 ft
* DS. RES. HEAD = 0 ft

| | Node # = | 1 | 2 | 3 | 4 | 5 |
|---|---|---|---|---|---|---|
| $t = 0$ | $Q_I =$ | 11.0 | 11.0 | 11.0 | 11.0 | 11.0 |
| $C_d = 0.24$ | $H_I =$ | 50.0 | 50.0 | 50.0 | 50.0 | 50.0 |
| $t = 0.33$ | $Q_I =$ | 11.0 | 11.0 | 11.0 | 11.0 | 0.0 |
| $C_d = 0$ | $H_I =$ | 50.0 | 50.0 | 50.0 | 50.0 | 703.5 |
| $t = 0.67$ | $Q_I =$ | 11.0 | 11.0 | 11.0 | 0.0 | 0.0 |
| $C_d = 0$ | $H_I =$ | 50.0 | 50.0 | 50.0 | 703.5 | 703.5 |
| $t = 1.00$ | $Q_I =$ | 11.0 | 11.0 | 0.0 | 0.0 | 0.0 |
| $C_d = 0$ | $H_I =$ | 50.0 | 50.0 | 703.5 | 703.5 | 703.5 |
| $t = 1.33$ | $Q_I =$ | 11.0 | 0.0 | 0.0 | 0.0 | 0.0 |
| $C_d = 0$ | $H_I =$ | 50.0 | 703.5 | 703.5 | 703.5 | 703.5 |
| $t = 1.67$ | $Q_I =$ | −11.0 | 0.0 | 0.0 | 0.0 | 0.0 |
| $C_d = 0$ | $H_I =$ | 50.0 | 703.5 | 703.5 | 703.5 | 703.5 |
| $t = 2.00$ | $Q_I =$ | −11.0 | −11.0 | 0.0 | 0.0 | 0.0 |
| $C_d = 0$ | $H_I =$ | 50.0 | 50.0 | 703.5 | 703.5 | 703.5 |
| $t = 2.33$ | $Q_I =$ | −11.0 | −11.0 | −11.0 | 0.0 | 0.0 |
| $C_d = 0$ | $H_I =$ | 50.0 | 50.0 | 50.0 | 703.5 | 703.5 |
| $t = 2.67$ | $Q_I =$ | −11.0 | −11.0 | −11.0 | −11.0 | 0.0 |
| $C_d = 0$ | $H_I =$ | 50.0 | 50.0 | 50.0 | 50.0 | 703.5 |
| $t = 3.00$ | $Q_I =$ | −11.0 | −11.0 | −11.0 | −11.0 | 0.0 |
| $C_d = 0$ | $H_I =$ | 50.0 | 50.0 | 50.0 | 50.0 | −603.5 |

Table 9.2 lists data for instant valve closure in the same pipe with friction (Case b). The difference in these data, compared with those in table 9.1, is that the head at the valve continues to rise after the valve is closed. This is caused by line packing, discussed in Section 8.5. Attenuation can also be observed. At $t = 1.333$ s, the wave is at node 2 and the head rise is $328 - 40 = 287$ ft, which is less than the initial head rise of 313 ft at node 5 at

**TABLE 9.2   Data for Example 9.2, Case b**
   *Input Data*

* PIPE LENGTH = 2,000 ft
* NUMBER OF PIPE SECTIONS = 4
* PIPE DIA. = 1.00 ft
* VALVE DIA. = 1.00 ft
* FRICTION FACTOR = 0.030
* WAVE SPEED = 1500 fps
* VALVE OPER. TIME = 0 s
* PROG. RUN TIME = 3.0 s
* US. RES. HEAD = 50 ft
* DS. RES. HEAD = 0 ft

| | Node # = | 1 | 2 | 3 | 4 | 5 |
|---|---|---|---|---|---|---|
| $t = 0$ | $Q_I =$ | 5.1 | 5.1 | 5.1 | 5.1 | 5.1 |
| $C_d = 0.24$ | $H_I =$ | 50.0 | 40.2 | 30.4 | 20.5 | 10.7 |
| $t = 0.33$ | $Q_I =$ | 5.1 | 5.1 | 5.1 | 5.1 | 0.0 |
| $C_d = 0$ | $H_I =$ | 50.0 | 40.2 | 30.4 | 20.5 | 313.2 |
| $t = 0.67$ | $Q_I =$ | 5.1 | 5.1 | 5.1 | 0.1 | 0.0 |
| $C_d = 0$ | $H_I =$ | 50.0 | 40.2 | 30.4 | 318.1 | 313.2 |
| $t = 1.00$ | $Q_I =$ | 5.1 | 5.1 | 0.2 | 0.1 | 0.0 |
| $C_d = 0$ | $H_I =$ | 50.0 | 40.2 | 323.0 | 318.1 | 323.0 |
| $t = 1.33$ | $Q_I =$ | 5.1 | 0.2 | 0.2 | 0.1 | 0.0 |
| $C_d = 0$ | $H_I =$ | 50.0 | 328.0 | 323.0 | 327.9 | 323.0 |
| $t = 1.67$ | $Q_I =$ | −4.4 | 0.2 | 0.2 | 0.1 | 0.0 |
| $C_d = 0$ | $H_I =$ | 50.0 | 328.0 | 332.8 | 327.9 | 332.8 |
| $t = 2.00$ | $Q_I =$ | −4.4 | −4.5 | 0.2 | 0.1 | 0.0 |
| $C_d = 0$ | $H_I =$ | 50.0 | 58.6 | 332.8 | 337.7 | 332.8 |
| $t = 2.33$ | $Q_I =$ | −4.5 | −4.5 | −4.5 | 0.1 | 0.0 |
| $C_d = 0$ | $H_I =$ | 50.0 | 58.6 | 67.3 | 337.7 | 342.6 |
| $t = 2.67$ | $Q_I =$ | −4.5 | −4.5 | −4.5 | −4.5 | 0.0 |
| $C_d = 0$ | $H_I =$ | 50.0 | 58.7 | 67.3 | 76.0 | 342.6 |
| $t = 3.00$ | $Q_I =$ | −4.5 | −4.5 | −4.5 | −4.5 | 0.0 |
| $C_d = 0$ | $H_I =$ | 50.0 | 58.7 | 67.3 | 76.0 | −183.1 |

$t = 0.333$. Also notice that there is still some flow at node 2 (0.2 cfs). The maximum head (343 ft), which occurs at $t = 2.33$ s, is less than $aV_o/g + HR_1 = 353$ ft because of attenuation caused by friction losses. Note in Table 9.2 that the head at the valve stays constant for two time steps. This is because of the first-order friction assumption where the friction loss is based on the flow at the adjacent node at the previous time step. For example, for

**TABLE 9.3   Data for Example 9.2, Case C**

*Input Data*
* PIPE LENGTH = 2,000 ft
* NUMBER OF PIPE SECTIONS = 4
* PIPE DIA. = 1.00 ft
* VALVE DIA. = 1.00 ft
* FRICTION FACTOR = 0.3
* WAVE SPEED = 1500 fps
* VALVE OPER. TIME = 2.67 s
* PROG. RUN TIME = 5.00 s
* US. RES. HEAD = 50 ft
* DS. RES. HEAD = 0 ft

| Node # = | | 1 | 2 | 3 | 4 | 5 |
|---|---|---|---|---|---|---|
| $t = 0$ | $Q_I =$ | 5.1 | 5.1 | 5.1 | 5.1 | 5.1 |
| $C_d = 0.240$ | $H_I =$ | 50.0 | 40.2 | 30.4 | 20.5 | 10.7 |
| $t = 0.33$ | $Q_I =$ | 5.1 | 5.1 | 5.1 | 5.1 | 5.0 |
| $C_d = 0.190$ | $H_I =$ | 50.0 | 40.2 | 30.4 | 20.5 | 16.8 |
| $t = 0.67$ | $Q_I =$ | 5.1 | 5.1 | 5.1 | 5.0 | 4.8 |
| $C_d = 0.145$ | $H_I =$ | 50.0 | 40.2 | 30.4 | 26.4 | 27.2 |
| $t = 1.00$ | $Q_I =$ | 5.1 | 5.1 | 5.0 | 4.8 | 4.6 |
| $C_d = 0.115$ | $H_I =$ | 50.0 | 40.2 | 36.0 | 36.5 | 40.0 |
| $t = 1.33$ | $Q_I =$ | 5.1 | 5.0 | 4.8 | 4.6 | 4.2 |
| $C_d = 0.085$ | $H_I =$ | 50.0 | 45.7 | 45.9 | 48.9 | 62.4 |
| $t = 1.67$ | $Q_I =$ | 4.9 | 4.8 | 4.6 | 4.3 | 3.8 |
| $C_d = 0.065$ | $H_I =$ | 50.0 | 55.2 | 57.8 | 70.7 | 87.5 |
| $t = 2.00$ | $Q_I =$ | 4.6 | 4.6 | 4.3 | 3.9 | 2.7 |
| $C_d = 0.035$ | $H_I =$ | 50.0 | 61.7 | 79.0 | 95.2 | 154.3 |
| $t = 2.33$ | $Q_I =$ | 4.2 | 4.1 | 3.8 | 2.8 | 1.2 |
| $C_d = 0.013$ | $H_I =$ | 50.0 | 73.3 | 97.9 | 160.6 | 246.0 |
| $t = 2.67$ | $Q_I =$ | 3.6 | 3.5 | 2.6 | 1.2 | 0.0 |
| $C_d = 0$ | $H_I =$ | 50.0 | 85.6 | 153.2 | 246.3 | 323.8 |
| $t = 3.00$ | $Q_I =$ | 2.8 | 2.2 | 1.0 | −0.1 | 0.0 |
| $C_d = 0$ | $H_I =$ | 50.0 | 128.8 | 231.9 | 315.1 | 318.7 |
| $t = 3.33$ | $Q_I =$ | 0.8 | 0.4 | −0.6 | −0.2 | 0.0 |
| $C_d = 0$ | $H_I =$ | 50.0 | 195.0 | 289.8 | 304.2 | 306.5 |
| $t = 3.67$ | $Q_I =$ | −2.1 | −1.9 | −0.9 | −0.4 | 0.0 |
| $C_d = 0$ | $H_I =$ | 50.0 | 210.8 | 267.2 | 281.3 | 289.7 |
| $t = 4.00$ | $Q_I =$ | −4.6 | −3.3 | −1.7 | −0.6 | 0.0 |
| $C_d = 0$ | $H_I =$ | 50.0 | 122.9 | 202.9 | 252.9 | 256.1 |
| $t = 4.33$ | $Q_I =$ | 4.5 | −4.4 | −3.0 | −1.3 | 0.0 |
| $C_d = 0$ | $H_I =$ | 50.0 | 45.5 | 110.5 | 178.3 | 216.2 |
| $t = 4.67$ | $Q_I =$ | −4.2 | −4.2 | −3.9 | −2.4 | 0.0 |
| $C_d = 0$ | $H_I =$ | 50.0 | 39.6 | 24.2 | 75.6 | 101.2 |
| $t = 5.00$ | $Q_I =$ | −3.9 | −3.7 | −3.5 | −2.5 | 0.0 |
| $C_d = 0$ | $H_I =$ | 50.0 | 29.1 | 6.9 | −50.1 | −62.9 |

node 5 (valve) at $t = 0.333$, friction is based on $Q_4 = 5.10$ at $t = 0$. At the same node at $t = 0.667$, friction is based on $Q_4 = 5.10$ at $t = 0.333$. Since the flow did not change, no change occurs in the friction, so $H_5$ stays constant for two time steps. At $t = 1.00$, $Q_4 = 0$, so friction goes to zero and $HP_5$ increases. The process repeats each two time steps. This is less of a problem for slow valve closure and for a larger number of pipe sections.

Table 9.3 shows results for the same system but for a closure in $2L/a$ (Case c). Each time the valve is closed an increment, a pressure wave is generated and transmitted toward the reservoir. Compare this with Fig. 8.4. The maximum head is less than in Table 9.2. This is due to attenuation and the first-order friction approximation. Also, note how the negative waves reflected from the reservoir reduce the head at the valve after $2L/a$.

Example 9.2 demonstrated how the computer program properly simulates the movement of the pressure waves through the pipe and how it properly produces attenuation and line packing. The next several examples demonstrate how the program can be used to determine the safe closing time for a valve (Example 9.3) and to help select the size (Example 9.3) and type (Example 9.4) of valve best suited for transient control, and the influence of the wave speed (Example 9.6) transients caused by opening a valve (Example 9.7).

**Example 9.3.** Re-work Example 8.2 (Case b) to determine the safe closing time for the valve to limit the maximum pressure head to 288 ft (125 psi) in the pipe.

The system data for the example (see Example 4.1 and 8.2) are $D = 0.833$ ft, $fL/D = 250$, $XL = 17,000$ ft, so $f = 0.0123$, $RH_1 = 200$, $RH_2 = 0$ and $a = 1,500$ fps. Keep $N = 4$. Assume the pipe is horizontal with the datum through the pipe centerline so $z_1 = 0$ (at all nodes). Use a butterfly valve going from full open to full closed with $C_d$ values selected from Fig. 4.3.

Run the program for a range of $t_c$ values and list $H_{max}$ at the valve versus $t_c$. To reduce the amount of output and provide the $H_{max}$ values, a few modifications to the program would be helpful (but not necessary). Make the output of $Q_I$ and $H_I$ optional by use of a printing index, keep track of the maximum head $H_{max}$ at the valve with a simple if statement, and print out $H_{max}$.

The results for this example are shown below. The $t_c$ required to keep $H_{max}$ at 288 ft (125 psi) is about 190 s.

| $t_c$ sec | $H_{max}$ feet |
|---|---|
| 0 | 480 |
| 11.3 | 474 |
| 22.7 | 449 |
| 45.3 | 420 |
| 90.7 | 395 |
| 181 | 320 |
| 282 | 263 |
| 725 | 220 |

$C_d = 0.80, \ 0.64, \ 0.53, \ 0.47, \ 0.39, \ 0.30, \ 0.21,$
$0.14, \ 0.07, \ 0.02, \ 0.$

The data also demonstrate the effect of pipe length on the instant closure time as discussed in Section 9.4. For this problem, $2L/a = 22.7$ s. For $t_c = 0$, $H_{max} = 480$ ft. At $t_c = 45.4$ s, $H_{max} = 423$ ft. Thus the maximum head is only reduced 12% by increasing $t_c$ from 0 to 45 s. The head rise for $t_c = 2L/a$ is less than for $t_c = 0$ due to attenuation.

***Example 9.4.*** Evaluate the influence on the transient head rise of using a valve that is smaller than the pipe diameter. In Chapter 4 and Example 4.1, it was pointed out that for long pipelines it may be advisable to use a smaller valve to improve control and reduce the required valve closing time.

To use the basic waterhammer program for this example requires entering the valve diameter to calculate the valve area and using it for the calculations of the initial conditions and for the valve boundary. Use the pipeline in Ex 9.3 and try 10-, 6-, and 4-in. butterfly valves. The results for $H_{max}$ are

| Valve Diameter | QI (cfs) | $t_c = 45.3$ s (ft) | $t_c = 90.6$ s (ft) | $t_c = 181$ s (ft) | $t_c = 282$ s (ft) |
|---|---|---|---|---|---|
| 10 | 3.90 | 420 | 395 | 320 | 263 |
| 6 | 3.87 | 405 | 318 | 259 | 240 |
| 4 | 3.74 | 370 | 276 | 230 | 223 |

Because the friction loss is large compared with the full open valve loss, reducing the valve diameter to 4 in. only reduces the initial flow 4%. However, the maximum transient head rise is significantly reduced. This is because the effective closure time for the small valve is longer than for the large valve for the same total closure time. The 10-in. valve does not reduce the flow by 10% until it has closed about

70%. The 4-in. valve reduces the flow 10% after it has closed only 30%.

Improved control and a reduced transient head rise can also be achieved by proper valve selection. For a long pipeline where the velocities are low, Example 9.4 demonstrates how increasing the full open valve loss really has little influence on the initial flow rate. The effect of valve type is demonstrated in the next example.

**Example 9.5.** Evaluate the influence of valve type on transients by solving Example 4.1 for both Case A and Case B using a Howell–Bunger, cone, and globe valve (all 8 in.). Use the program listed in Fig. 9.3.

Use the following $C_d$ values at each 10% for the three valves:

| Valve Type | $C_d$ values | | | | | | |
|---|---|---|---|---|---|---|---|
| Howell–Bunger | 0.86, | 0.83, | 0.80, | 0.76, | 0.68, | 0.59, | 0.49, |
| | 0.38, | 0.26, | 0.13, | 0 | | | |
| Cone | 0.97, | 0.88, | 0.63, | 0.45, | 0.33, | 0.24, | 0.17, |
| | 0.105, | 0.055, | 0.03, | 0 | | | |
| Globe | 0.41, | 0.39, | 0.35, | 0.31, | 0.26, | 0.20, | 0.14, |
| | 0.090, | 0.040, | 0.01, | 0 | | | |

The $C_d$ values were taken from Figs. 4.3 and 4.4.
System data: $z_I = 0$ (all nodes), $N = 4$, $f = 0.0123$, $D = 0.833$ ft, $HR_2 = 0$, $a = 1,500$ fps

| | $XL$(ft) | $HR_1$(ft) | $t_c$(s.) |
|---|---|---|---|
| Case A | 200 | 30 | 2 |
| Case B | 17,000 | 200 | 100 |
| Results: | | | |

| | Case A | | Case B | |
|---|---|---|---|---|
| Valve type | $QI$ (cfs) | $H_{max}$ (ft) | $QI$ (cfs) | $H_{max}$ (ft) |
| Howell–Bunger | 13.2 | 704 | 3.90 | 429 |
| Cone | 13.8 | 209 | 3.90 | 367 |
| Globe | 8.5 | 177 | 3.87 | 370 |

For the short pipe (Case A), the type of valve used significantly influences the initial flow $QI$. For the long pipe, the valve type makes little difference.

The most dramatic effect of valve type on the transient head $H_{max}$ is for Case A between the Howell–Bunger and cone valves. Since the Howell–Bunger valve has larger values of $C_d$ at small angles, a much larger change of flow occurs during the last part of the valve movement, and this causes a larger head rise.

***Example 9.6.*** Evaluate the influence of wave speed on the transient head rise. Use the pipe in Example 9.3 to find $H_{max}$ for $a = 1500$ and 3000 fps. The results for $a = 1500$ fps from Example 9.3 are listed below with the new data for $a = 3000$ fps.

| $t_c$ | $H_{max}$ (1500 fps) | $H_{max}$ (3000 fps) |
|-------|----------------------|----------------------|
| 0     | 480                  | 815                  |
| 11.3  | 474                  | 772                  |
| 22.7  | 449                  | 740                  |
| 45.3  | 420                  | 690                  |
| 90.7  | 395                  | 470                  |
| 181   | 320                  | 309                  |
| 282   | 263                  | 264                  |
| 725   | 220                  | 218                  |

$C_d = 0.80, 0.64, 0.53, 0.47, 0.39, 0.30, 0.21, 0.14, 0.05, 0.02, 0$

At $t_c = 0$, the head is not quite doubled by doubling the wave speed. Again, this is due to friction. Also note that the influence of wave speed disappears as $t_c$ increases. This is because the slower reflections from the reservoir for the smaller wave speed offset the smaller incremental transient head rise.

***Example 9.7.*** Calculate the transient generated by instantly opening a valve in the system shown in Fig. 9.4. Use the program listed in Fig. 9.3 with the necessary modifications.

The first program modification is to calculate the initial conditions so $Q_I = 0$, $H_I = HR_1$ at all nodes. The second modification is to enter $C_{dI}$ from closed to open for $t > t_c$. For $t > t_c$, set $CC = C_{d11}$ which will be the final valve opening. Also change the calculation of $HP_{NS}$ and $QP_{NS}$ from that in lines 45–46 of Fig. 9.3. Note that a $C_d$ value of 1.0 causes C3 (line 39 in Fig. 9.3) to become zero. To avoid this do not allow $C_d$ to equal 1.0.

The input data are printed at the top of Table 9.4. The results are listed in Table 9.4 and in Fig. 9.4. At $t = 0$, the data in Table 9.4 shows that $Q = 0$ and $H = HR_1 = 210$ ft at all five nodes. At $t = 1.667$ s, the valve is full open and the head at the valve drops to $HR_2$ (since

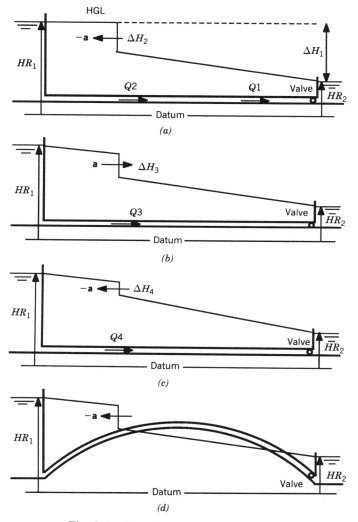

**Fig. 9.4**    Transient for valve opening.

$C_d = 0.9$ at full open there is negligible valve loss). The flow jumps to 1.69 cfs. This can be checked by solving Eq. 8.2 in the form $\Delta H = -a\Delta Q/gA = -B\Delta Q$ or $\Delta Q = -\Delta H/B$. $B = 118.6$ and $\Delta H = -200$, so $\Delta Q = 200/118.6 = 1.69$ cfs.

The negative pressure wave travels toward the upstream reservoir, as shown by the first diagonal line in Table 9.4. Note how the $Q$ at each upstream node is less and the head is higher than at the valve. This is caused by attenuation due to friction.

**TABLE 9.4** **Valve Opening Transient**

*Input Data*
* PIPE LENGTH = 20,000 ft
* NUMBER OF PIPE SECTIONS = 4
* PIPE DIA. = 1.00 ft
* VALVE DIA. = 1.00 ft
* FRICTION FACTOR = 0.020
* WAVE SPEED = 3000 fps
* VALVE OPER. TIME = 0 s
* PROG. RUN TIME = 30 s
* US. RES. HEAD = 210 ft
* DS. RES. HEAD = 10 ft

| Node # = | | 1 | 2 | 3 | 4 | 5 |
|---|---|---|---|---|---|---|
| $t = 0$ | $Q_I =$ | 0.0 | 0.0 | 0.0 | 0.0 | 0.0 |
| $C_d = 0$ | $H_I =$ | 210.0 | 210.0 | 210.0 | 210.0 | 210.0 |
| $t = 1.67$ | $Q_I =$ | 0.0 | 0.0 | 0.0 | 0.0 | 1.7 |
| $C_d = 0.9$ | $H_I =$ | 210.0 | 210.0 | 210.0 | 210.0 | 10.1 |
| $t = 3.33$ | $Q_I =$ | 0.0 | 0.0 | 0.0 | 1.7 | 1.7 |
| $C_d = 0.9$ | $H_I =$ | 210.0 | 210.0 | 210.0 | 13.6 | 10.1 |
| $t = 5.00$ | $Q_I =$ | 0.0 | 0.0 | 1.6 | 1.7 | 1.6 |
| $C_d = 0.9$ | $H_I =$ | 210.0 | 210.0 | 17.1 | 13.6 | 10.1 |
| $t = 6.67$ | $Q_I =$ | 0.0 | 1.6 | 1.6 | 1.6 | 1.6 |
| $C_d = 0.9$ | $H_I =$ | 210.0 | 20.4 | 17.1 | 13.6 | 10.1 |
| $t = 8.33$ | $Q_I =$ | 3.1 | 1.6 | 1.6 | 1.6 | 1.6 |
| $C_d = 0.9$ | $H_I =$ | 210.0 | 20.4 | 16.9 | 13.6 | 10.2 |
| $t = 10.00$ | $Q_I =$ | 3.1 | 3.0 | 1.6 | 1.5 | 1.6 |
| $C_d = 0.9$ | $H_I =$ | 210.0 | 197.2 | 16.9 | 13.5 | 10.2 |
| $t = 11.67$ | $Q_I =$ | 3.0 | 3.0 | 2.9 | 1.5 | 1.5 |
| $C_D = 0.9$ | $H_1 =$ | 210.00 | 197.2 | 185.2 | 13.5 | 10.1 |
| $t = 13.33$ | $Q_I =$ | 3.0 | 2.9 | 2.9 | 2.9 | 1.5 |
| $C_d = 0.9$ | $H_I =$ | 210.0 | 197.9 | 185.2 | 173.8 | 10.1 |
| $t = 15.00$ | $Q_I =$ | 2.8 | 2.9 | 2.8 | 2.9 | 4.1 |
| $C_d = 0.9$ | $H_I =$ | 210.0 | 197.9 | 186.5 | 173.8 | 10.1 |
| $t = 16.67$ | $Q_I =$ | 2.8 | 2.7 | 2.8 | 3.9 | 4.1 |
| $C_d = 0.9$ | $H_I =$ | 210.0 | 198.5 | 186.5 | 33.8 | 10.1 |
| $t = 18.33$ | $Q_I =$ | 2.7 | 2.7 | 3.8 | 3.9 | 3.8 |
| $C_d = 0.9$ | $H_I =$ | 210.0 | 198.5 | 55.7 | 33.8 | 10.2 |
| $t = 20.00$ | $Q_I =$ | 2.7 | 3.6 | 3.8 | 3.7 | 3.8 |
| $C_d = 0.9$ | $H_I =$ | 210.0 | 76.2 | 55.7 | 32.3 | 10.2 |
| $t = 21.67$ | $Q_I =$ | 4.5 | 3.6 | 3.6 | 3.7 | 3.6 |
| $C_d = 0.9$ | $H_I =$ | 210.0 | 76.2 | 53.0 | 32.3 | 10.1 |
| $t = 23.33$ | $Q_I =$ | 4.5 | 4.3 | 3.6 | 3.5 | 3.6 |
| $C_d = 0.9$ | $H_I =$ | 210.0 | 177.3 | 53.0 | 30.9 | 10.1 |
| $t = 25.00$ | $Q_I =$ | 4.2 | 4.3 | 4.2 | 3.5 | 3.4 |
| $C_d = 0.9$ | $H_I =$ | 210.0 | 177.3 | 146.8 | 30.9 | 10.0 |
| $t = 26.67$ | $Q_I =$ | 4.2 | 4.1 | 4.2 | 4.1 | 3.4 |
| $C_d = 0.9$ | $H_I =$ | 210.0 | 179.2 | 146.8 | 118.3 | 10.0 |
| $t = 28.33$ | $Q_I =$ | 4.0 | 4.1 | 4.0 | 4.1 | 4.6 |
| $C_d = 0.9$ | $H_I =$ | 210.0 | 179.2 | 150.4 | 118.3 | 10.2 |
| $t = 30.00$ | $Q_I =$ | 4.0 | 3.9 | 4.0 | 4.5 | 4.6 |
| $C_d = 0.9$ | $H_I =$ | 210.0 | 181.0 | 150.4 | 49.2 | 10.2 |

226

Figure 9.4a shows the hydraulic grade line at $t = 6.667$ s. The data are also listed in Table 9.4 for $t \sim 6.667$ s. To the right of the wave front there is flow, so the HGL slopes in the direction of flow. This makes $\Delta H_2 < \Delta H_1$ and by Eq. 8.1, $Q_2 < Q_1$. This is seen in Table 9.4.

When the wave hits the reservoir, the flow increases by a double increment, one increment by the negative wave and the other by the subsequent positive wave reflected from the reservoir. Table 9.4 shows that $Q_1 = 3.14$ cfs at $t = 8.33$ s.

The positive wave moves down the pipe along the second diagonal line in Table 9.4. Figure 9.4b shows the HGL at $t \sim 10.0$ s. Since $\Delta H_3 < \Delta H_2$, the flow increases by a little less than double $Q_2$, so $Q_3 = Q_2 + \Delta H_3 / B$.

The wave continues to the downstream reservoir and is reflected. Figure 9.4c shows the HGL at $t \sim 20$ s. The HGL is steeper since the flow has increased, $\Delta H_4 < \Delta H_3$ and $Q_4 > Q_3$. The process continues until the wave dampens out by friction and $Q$ reaches steady state.

Note that the maximum transient does not exceed the initial reservoir head, so the transient caused by opening a valve too quickly does not cause any problem for a horizontal pipeline. Friction rapidly dampens the transient and the flow accelerates gradually. This problem is also solved with surge analysis in Section 9.9.

Figure 9.4d shows a pipe with a high point where the pressure during a valve opening transient can cause negative pressure in the pipe. This could result in pipe collapse, drawing in of contaminated groundwater (if there are leaks), opening of air-vacuum valves, or column separation, all of which are undesirable and should be avoided. Therefore, the opening time of valves can be important.

## 9.8 ADDITIONAL BOUNDARY CONDITIONS

### Orifice at End of Pipe

This boundary condition is similar to the valve at the end of a pipe. It produces the same quadratic (Eq. 9.33). The only difference is that $K_l$ is a constant.

### Dead End Downstream

This is an important boundary condition for a branching pipe system where one line may be closed off. At the closed end, $QP_{NS} = 0$. The head is found from the proper compatibility equation (9.26 or 9.27).

### Outlet Valve at Interior Point

Consider an outlet valve located at an interior node $NV$ discharging into a small downstream reservoir, as shown in Fig. 9.5. In analyzing this boundary condition, it is assumed that 1) the distance from the main pipe to the discharge tank is negligible, 2) there is no pressure drop at the connection so there is a common head in all pipes, 3) the piezometric head $HR_2$ in the discharge tank is constant, and 4) the exit loss into the tank is not included.

The unknowns are $QP_{NV}$, $QPD$, $QPV$, and $HP_{NV}$. Therefore, four equations will be required. These will be the $C^+$ and $C^-$ compatibility equations, the continuity equation, and the equation for head drop across the valve.

$$C^+: HP_{NV} = CP - B\,QP_{NV} \tag{9.34}$$

$$C^-: HP_{NV} = CM + B\,QPD \tag{9.35}$$

$$\text{Continuity } QP_{NV} = QPV + QPD \tag{9.36}$$

$$\text{Valve: } HP_{NV} = HR_2 + C3\,QPV^2 \tag{9.37}$$

$$C3 = \pm K_l/(2gA_v^2)$$

Solve Eqs. 9.34 and 9.35 for $QP$ and $QPD$ then substitute them into Eq. 9.36 and rearrange:

$$QPV = QP_{NV} - QPD = \frac{[CP - HP_{NV}]}{B} - \frac{[HP_{NV} - CM]}{B}$$

$$= \frac{[CP + CM - 2HP_{NV}]}{B}$$

Next substitute this into Eq. 9.37. After simplifying,

$$HP_{NV}^2 + C4\,HP_{NV} + C5 = 0$$

in which

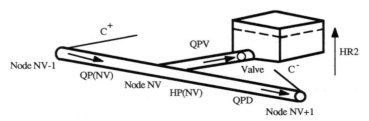

**Fig. 9.5** Outlet valve at interior point.

$$C4 = -CP - CM - B^2/(4C3)$$

$$C5 = 0.25(CP + CM)^2 + B^2 HR_2/(4C3)$$

so

$$HP_{NV} = 0.5(-C4 \pm \sqrt{C4^2 - 4C5}) \qquad (9.38)$$

In determining the sign of the square root for positive flow, consider frictionless steady state conditions where $H_{I-1} = H_{I+1}$. Remember that the analysis must be valid for steady as well as unsteady flow. For this case, $-CP - CM = -2H$ (see Eqs. 9.28 and 9.29). $B^2/4C3$ is always positive, so $C4$ must be negative. This makes $-C4$ a positive quantity and larger than $HP_{NV}$. The only way to have the right side of Eq. 9.38 satisfy the required conditions for steady state and equal $HP_{NV}$ is to use the negative sign. The solution is completed by solving Eqs. 9.34–9.36 for the other three unknowns. For reverse flow from the tank, use $-C3$ and the positive sign in Eq 9.38.

When incorporating this into a program, the procedure for handling the $K_l$ values of the valve as it opens or closes is the same as used in the program in Fig. 9.3.

**Minor loss**

Most minor (or local) losses can be ignored unless they cause a significant head loss. The pressure wave passes through elbows, large orifices, full open valves, etc., with little attenuation or reflection. Pipe entrances and exits generally have negligible losses. For cases where the loss may be significant, the following solution for a minor loss in a pipe of constant diameter may be used.

With a fixed loss at an interior node $NL$ as shown in Fig. 9.6, there are three unknowns: $QP_{NL}$, $HP_{NL}$ (upstream pressure), and $HPD$ (head downstream). Three equations are needed:

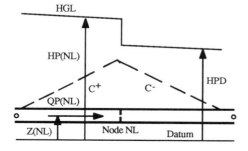

**Fig. 9.6**  Minor loss at interior node.

$$C^+: HP_{NL} = CP - B\,QP_{NL} \qquad\qquad (9.39)$$

$$C^-: HPD = CM + B\,QP_{NL} \qquad\qquad (9.40)$$

$$HP_{NL} - HPD = C3\,QP_{NL}^2, \qquad C3 = \pm\frac{K_l}{(2gA_v^2)} \qquad (9.41)$$

Substituting Eqs. 9.39 and 9.40 into 9.41 and solving gives

$$QP_{NL} = 0.5(-C4 \pm \sqrt{C4^2 - 4C5}) \qquad\qquad (9.42)$$

in which $C4 = 2B/C3$ and $C5 = (CM - CP)/C3$. Use the positive sign for forward flow and the negative sign for reverse flow. The two heads are found from eqs 9.39 and 9.40.

### In-line Valve

The solution for a valve installed at an interior point is solved with the equations developed for a minor loss. The only difference is that $K_l$ changes as the valve closes. The procedure outlined in the basic program contained in Fig. 9.3 can be used to obtain the $K_l$ values.

### Centrifugal Pump

The steady-state head–discharge characteristics of centrifugal pumps is shown in Fig. 3.3. When, operating near their design point, the curve can be expressed with Eq. 3.3. Figure 9.7 shows the steady-state pump boundary condition. The problem is simplified by ignoring the velocity head. This makes the total pump head $H_p$ equal to the pressure increase at the pump or $H_p = HGL - HR_1$, and $H_p = H_o - C1\,QP - C2\,QP^2$. Putting Eq. 3.3 in terms of the piezometric head $HP_1$ and $QP_1$ gives

$$HP_1 = H_o - C1\,QP - C2\,QP^2 + HR_1 \qquad\qquad (9.43)$$

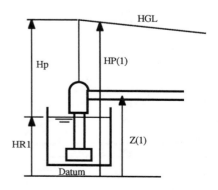

**Fig. 9.7**  Pump at upstream node.

If the pump is at the upstream boundary, $HP_1$ and $QP_1$ are found by solving Eq. 9.43 with the $C^-$ compatibility equation (Eq. 9.27). The solution is

$$QP_1 = 0.5(-C4 + \sqrt{C4^2 - 4C5})  \qquad (9.44)$$

in which $C4 = (C1 + B)/C2$ and $C5 = (CM - H_o - HR_1)/C2$. The head is found using Eq. 9.43. Since $C4$ is positive, the positive sign is used to get positive flow. These equations are only good for normal pump operation, not for pump failures. See Reference 87 for further information regarding pump failures.

## 9.9   SURGE ANALYSIS

As discussed in Section 9.1, there are unsteady flows where the analysis can be done assuming rigid column motion. Examples are starting a pump, opening a valve, an oscillating $U$-tube, and oscillation of a surge tank. The solution involves applying the equation of motion $F = m \, dV/dt$ to the mass of water. For frictionless systems, an analytical solution is possible. With friction, a numerical solution is needed.

**Example 9.8.** Apply surge analysis to the system shown in Fig. 9.4a to calculate the time required to establish steady-state flow. The forces acting on the water in the pipe are pressure, gravity, and friction. If the piezometric head is used in the following derivation the solution is valid for either a horizontal or inclined pipe. The valve is assumed to have no loss fully open.

The equation of motion, applied to Fig. 9.4a, is

$$\gamma(HR_1 - HR_2)A - \frac{\gamma AfLQ|Q|}{2gDA^2} = \frac{\gamma L}{g}\frac{dQ}{dt}  \qquad (9.45)$$

Using a first-order friction assumption ($Q = Q1$), dividing by $\gamma A$, and putting the equation in finite difference form gives

$$HR_1 - HR_2 - \frac{fLQ1|Q1|}{2gDA^2} = \frac{L}{gA\Delta t}(Q2 - Q1)$$

$$C1 = HR_1 - HR_2 - \frac{fLQ1|Q1|}{2gDA^2}$$

and $C2 = L/(gA\Delta t)$, so $C1 = C2(Q2 - Q1)$ or $Q2 = Q1 + C1/C2$

$$\qquad (9.46)$$

An important limitation in solving Eq. 9.46 is to use a small enough time step that the change in $Q$ is small so that the first-order friction assumption is valid. Table 9.5 lists the output of a simple spreadsheet application for the problem in Example 9.7. Compare this output with that of the transient program in Table 9.4. The major difference is in the first 20 s, where the transient program accurately shows that there is a significant variation in the discharge along the pipe. After two cycles of the wave $(t > 4L/\mathbf{a})$, the friction has almost dampened out the transient, and both programs provide similar results.

Applying the energy equation to the system gives a steady-state flow of 4.46 cfs. At 52 s, the flow is exactly this value. This same type of analysis can be used for oscillations in a surge tank.

**TABLE 9.5   Spreadsheet Application for Example 9.7**

| Input Data | Assumed Data | | | Calculated Data | | |
|---|---|---|---|---|---|---|
| | Q1 | DT | T | C1 | C2 | Q2 |
| f = 0.02 | 0 | 0.2 | 0 | 200.0 | 3105.6 | 0.06 |
| D = 1 | 0.0644 | 0.4 | 0.4 | 200.0 | 1552.8 | 0.19 |
| L = 20,000 | 0.1932 | 0.6 | 1 | 199.6 | 1035.2 | 0.39 |
| g = 32.2 | 0.386 | 1 | 2 | 198.5 | 621.1 | 0.71 |
| HR1 = 210 | 0.7056 | 1 | 3 | 195.0 | 621.1 | 1.02 |
| HR2 = 10 | 1.0195 | 1 | 4 | 189.5 | 621.1 | 1.32 |
| A = 0.7854 | 1.3247 | 1 | 5 | 182.3 | 621.1 | 1.62 |
| | 1.6182 | 1 | 6 | 173.6 | 621.1 | 1.90 |
| | 1.8978 | 1 | 7 | 163.7 | 621.1 | 2.16 |
| | 2.1614 | 1 | 8 | 153.0 | 621.1 | 2.41 |
| | 2.4077 | 1 | 9 | 141.6 | 621.1 | 2.64 |
| | 2.6357 | 1 | 10 | 130.1 | 621.1 | 2.85 |
| | 2.8451 | 1 | 11 | 118.5 | 621.1 | 3.04 |
| | 3.0358 | 1 | 12 | 107.2 | 621.1 | 3.21 |
| | 3.2084 | 1 | 13 | 96.3 | 621.1 | 3.36 |
| | 3.3636 | 1 | 14 | 86.1 | 621.1 | 3.50 |
| | 3.5021 | 2 | 16 | 76.5 | 310.6 | 3.75 |
| | 3.7485 | 2 | 18 | 58.5 | 310.6 | 3.94 |
| | 3.9369 | 2 | 20 | 43.9 | 310.6 | 4.08 |
| | 4.0784 | 2 | 22 | 32.5 | 310.6 | 4.18 |
| | 4.1831 | 2 | 24 | 23.8 | 310.6 | 4.26 |
| | 4.2598 | 2 | 26 | 17.3 | 310.6 | 4.32 |
| | 4.3154 | 2 | 28 | 12.5 | 310.6 | 4.36 |
| | 4.3556 | 3 | 31 | 9.0 | 207.0 | 4.40 |
| | 4.399 | 3 | 34 | 5.2 | 207.0 | 4.42 |
| | 4.4239 | 3 | 37 | 2.9 | 207.0 | 4.44 |
| | 4.4381 | 4 | 41 | 1.7 | 155.3 | 4.45 |
| | 4.4488 | 5 | 46 | 0.7 | 124.2 | 4.45 |
| | 4.4546 | 6 | 52 | 0.2 | 103.5 | 4.46 |

## 9.10  CONTROLLING TRANSIENTS

### Pressure-Relief Valves

There are situations where transients can be effectively controlled with pressure-relief valves (PRVs). One example is the rapid closure of turbine wicket gates caused by load rejection. The pressure rise in the penstock can be controlled by simultaneously opening a PRV as the wicket gates close.

A PRV can also be placed at the discharge side of pumps to control transients caused by pump failure. A common application is as a "surge-anticipator valve," where the PRV is set to open in the event of a power failure during pumping. Power failure to a pump causes the discharge head to drop rapidly, the flow to stop and reverse, closing a check valve at the pump, and potentially a high pressure rise. Opening of the surge-anticipator valve controls the pressure in the pipe.

Another application of the PRV is to protect the pipeline from over-pressurization caused by starting a pump. It is placed in the pump discharge line between the pump and discharge control valve and functions as a bypass valve during pump start-up. When a pump is started and the pipeline is empty or only partly full, it is often necessary to operate the pump with the discharge valve part closed to fill the line slowly. To prevent a pump from operating near shutoff head for extended periods of time, the PRV can be set to open at a desired pressure and bypass flow back to the sump or into the suction side of the pump.

If the PRV is placed at an interior point, the boundary condition is similar to that of the outlet valve at an interior point. The functional difference is the manner in which the valve is opened. Relief valves usually only activate if the line pressure exceeds some set pressure.

There are two other factors affecting the solution of the PRV boundary condition. First is choking cavitation. PRVs normally operate with high upstream pressures and discharge through a short pipe into the atmosphere. This makes them susceptible to choking. Since choking increases the valve loss coefficient $K_l$, the input data for the valve must be corrected (see Chapter 6 for details).

With high velocity flow through the PRV, minor losses in the PRV line can be significant. These can be included easily by merely adding them to the $K_l$ values of the valve. With choking and minor losses properly handled, the boundary condition can be solved with Eqs. 9.38 and 9.34–9.36 if the PRV is at an interior point.

There are still some practical problems associated with the PRV boundary condition. The valve should not be oversized. For the PRV to operate satisfactorily, it should open quickly to relieve the pressure and close slowly so as not to generate its own transient. Obtaining complete information from valve suppliers on the dynamic behavior of their valves is difficult.

If a system is subjected to high positive transients, it is likely that it will also be subjected to low pressures, possible even vapor pressure. If the pressure in the pipe at a discharging PRV suddenly becomes subatmospheric, the valve may draw air into the pipe before it can close. From that time on, the program will no longer accurately simulate the boundary condition without modifications.

### Simple Surge Tanks

A simple surge tank (Fig. 9.8) is a vertical tank connected to the pipeline so as to minimize entrance losses. It has an open water surface, so it normally extends above the maximum hydraulic grade line. Its diameter is considerably larger than that of the pipe to avoid spilling. If spilling can be tolerated, the surge tank diameter can be close to the pipe diameter. Smaller diameter tanks may result in higher pressures in the pipe and therefore provide less protection. Small surge tanks will be referred to as standpipes and are discussed separately.

Simple surge tanks prevent transients generated in one section of pipe from being transmitted into another section. A typical installation would be to place the surge tank between the power tunnel and penstock of a power plant. It keeps most of the transient generated in the penstock from entering the power tunnel. During start-up of the turbine, the surge tank also supplies water while the water in the power tunnel is accelerating.

Figure 9.8 shows a simple surge tank located at an interior node $NT$. The derivation of the boundary condition is based on the following assumptions: 1) there is no loss at the inlet to the tank, 2) in a computing time interval, there is only a small change in the water level in the tank, 3) the tank diameter is large compared with that of the pipe, so that the pressure inside is hydrostatic and inertia forces can be ignored.

There are five unknowns identified in Fig. 9.8: $QP_{NT}$, $QPT$, $QPD$, $HP_{NT}$, and $XLPT$. The same quantities calculated for steady state, or known from

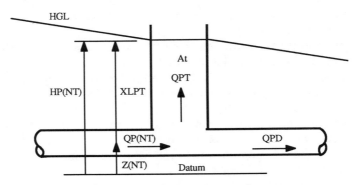

**Fig. 9.8**  Simple surge tank.

the previous time step, are $Q_{NT}$, $QT$, $QD$, $H_{NT}$, and $XLT$. The five equations are the $C^+$ and $C^-$ equations, continuity, hydrostatic pressure, and change of level in the tank.

$$HP_{NT} = CP - B\,QP_{NT} \tag{9.47}$$

$$HP_{NT} = CM + B\,QPD \tag{9.48}$$

$$QP_{NT} = QPT + QPD \tag{9.49}$$

$$HP_{NT} = XLPT + Z_{NT} \tag{9.50}$$

$$XLPT = XLT + QT\,\frac{\Delta t}{At} \tag{9.51}$$

Equation 9.51 is a first order approximation. As a result, solution of the five equations is direct starting with $XLPT$. The second order approximation of the change in tank level is

$$XLPT = XLT + \frac{\Delta t(QPT + QT)}{(2A_t)} \tag{9.52}$$

which requires a simultaneous solution. First, substitute Eq. 9.52, 9.47, and 9.49 into Eq. 9.50. Next, solve 9.47 and 9.48 for $QPD$ and substitute into the modified Eq. 9.50. The result is

$$QP_{NT} = (C1 + C3)/(2 + C2) \tag{9.53}$$

in which

$$C1 = 2At[CP - Z(NT) - XLT]/\Delta t - QT$$

$$C2 = 2AtB/\Delta t$$

$$C3 = (CP - CM)/B$$

$HP_{NT}$ is found with Eq. 9.47, $QPD$ from Eq. 9.48, $QPT$ from Eq. 9.49, and $XLPT$ from Eq. 9.50 or 9.52.

The accuracy of the solution is improved by using a smaller $\Delta t$. The accuracy can be checked by varying the number of sections $N$ (to change $\Delta t$) and compare results. The second order solution is recommended.

## Throttled Surge Tank

To reduce the size of a surge tank, a restriction can be placed at the inlet. The penalty for doing this is that the pressure is higher in the pipe and more of the transient continues past the tank into the upstream pipe. The amount of restriction is based on an economic and hydraulic analysis of how much the transient must be reduced. As the size and therefore the cost of the surge tank, are reduced, the pressure class (and cost) of the pipe may increase.

Figure 9.9 identifies the five unknowns. The figure depicts flow into the tank where the HGL is above the water level in the tank. It is assumed that the pressure in the tank is hydrostatic so the inertia forces required to accelerate the water in the tank are negligible. Using a first-order approximation for $XLPT$, the five equations are Eqs. 9.47–9.49 and 9.51 plus

$$HP_{NT} - XLPT - Z_{NT} = \pm C3 \, QPT^2 \qquad (9.54)$$

in which $C3 = \pm K_l/(2gA^2)$, $K_l$ being the loss coefficient for the surge tank entrance based on $A$ = area of pipe (the tank area could also be used if the appropriate $K_l$ is selected). The $+C3$ is used for flow into the tank and $-C3$ for flow out.

First solve Eq. 9.51 for $XLPT$. Then substitute Eqs. 9.47 and 9.48 into Eq. 9.49 and solve for $QPT$. Substituting this into Eq. 9.54 gives

$$HP_{NT} = 0.5(-C4 \pm \sqrt{C4^2 - 4 \, C5}) \qquad (9.55)$$

in which

$$C4 = -CP - CM \pm B^2/(4 \, C3) \qquad (9.56)$$

$$C5 = (CP + CM)^2/4 \pm \frac{B^2[XLT + Z_{NT}]}{(4 \, C3)} \qquad (9.57)$$

For inward flow use, the negative sign in Eqs. 9.55 and 9.56 and the positive sign in Eq. 9.57. For outward flow, use the opposite signs. One must also be concerned if the loss coefficient $K_l$ is different for inward and outward flow.

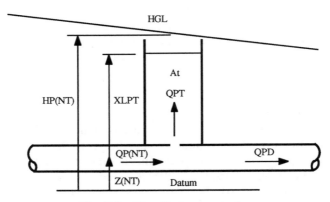

**Fig. 9.9**  Throttled surge tank.

## Standpipes

Some reduction in cost may result by using a small diameter surge tank if spillage can be allowed. In such a case, the tank acts much like a pressure-relief valve. The tank diameter can be less than the pipe diameter and the height reduced. It may be desirable to have it high enough to prevent spillage during a normal shutdown.

For the small-diameter tank, inertia forces may be required to accelerate the fluid in the tank. This makes the pressure at the bottom of the tank greater than hydrostatic pressure. The diameter of a standpipe will normally be close to the pipe diameter. If there is no restriction at the tank entrance, the loss will be small compared with the inertia force and may be ignored. Also, friction in the surge tank can be ignored. Based on these assumptions, the equations required to solve a standpipe located at an interior node $NT$ are $C^+$, $C^-$, continuity at the junction (Eqs. 9.47–9.49), a second-order estimate for the change of water level (Eq. 9.52), and the equation of motion ($\Sigma F = ma$). Applied to Fig. 9.9, the equation of motion is

$$[HP(NT) - Z_{NT} - XLPT]\gamma At = XLPTA_t\rho\Delta Q/(At\,\Delta t)$$

Divide by $\gamma At$ and let $\Delta Q = QPT - QT$:

$$HP_{NT} - Z_{NT} - XLPT = XLPT(QPT - QT)/(g\,\Delta t\,At) \qquad (9.58)$$

Solving Eq. 9.58 with Eqs. 9.47–9.49 and 9.52 gives

$$HP_{NT} = 0.5(-C7 \pm \sqrt{C7^2 - 4\,C8}) \qquad (9.59)$$

in which

$$C1 = XLT + \frac{\Delta t}{2At}\left[\frac{(CP + CM)}{B} + QT\right]$$

$$C2 = \frac{\Delta t}{(BA_t)}$$

$$C3 = g\,\Delta t\,A_t$$

$$C4 = \frac{(CP + CM)}{B} - QT$$

$$C5 = \frac{C1}{C3}$$

$$C6 = \frac{C2}{C3}$$

$$C7 = \frac{-B\left(C4C6 + 2\dfrac{C5}{B} + C2 + 1\right)}{(2\,C6)}$$

$$C8 = \frac{B(C4C5 + Z_{NT} + C1)}{(2C6)}$$

$QP(NT)$ is found from Eq. 9.47, $QPD$ from Eq. 9.48, $QPT$ from Eq. 9.49, and $XLPT$ from Eq. 9.52. One additional constraint is required for this case: the tank height. Once the water level exceeds the tank height, spillage will occur. The program should keep track of the amount of spillage.

If higher pressures can be tolerated in the penstock, it may be possible to increase the entrance loss by having the connection to the penstock smaller than the diameter of the standpipe. This requires adding the entrance loss to the list of equations (Eq. 9.54). It also reduces the spillage.

The period of oscillation for the surge tank or penstock (ignoring friction) is (87):

$$T = 2\pi\sqrt{LA_t/gA} \qquad (9.60)$$

in which $L$ is the length of the water column (pipe and tank length), $At$ the area of the tank, and $A$ the pipe area.

### Air Chambers

One of the limitations of the surge tank is that it can only be used if the head is relatively small. For high head systems, the closed surge tank, or air chamber, can provide similar transient protection. The air chamber is a closed vessel filled with water and compressed air. One drawback is that a source of compressed air and various controls are required which make it more subject to mechanical failure.

The following is a brief description of the operation of an air chamber located near the discharge of a pump. When the pump is turned off and the pressure drops in the pipe, the compressed air supplies water from the tank. A check valve closes to prevent reverse flow through the pump. As the water leaves the tank, the air expands, the pressure drops, and the rate of flow decreases. Eventually, the water in the pipe comes to rest and reverses. Water flows back into the tank, compressing the air. The system oscillates and ultimately comes to rest. The period of the oscillation and the associated head fluctuation depends on the size of both the air chamber and the system. It is important that the tank be large enough that it never empties and allows the air to flow into the pipe.

Figure 9.10 shows an air chamber with the variables identified. The unknowns are: $QP_{NT}$, $HP_{NT}$, $QPT$, $QPD$, and $VOL$. The known flows and head from the previous time step are $Q_{NT}$, $QT$, $QD$, and $H_{NT}$. It is assumed that the drop of water level in the tank is small compared with the heads and it is ignored; the entrance and exit losses are included.

The equations needed are Eqs. 9.47–9.49, and 9.55, conservation of mass in the chamber

$$VOL = VOL - 0.5(QPT + QT)\cdot\Delta t \qquad (9.61)$$

and the reversible polytropic relation

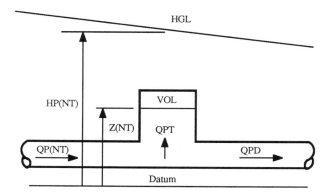

**Fig. 9.10**   Air chamber.

$$[HP_{NT} + H_b - Z_{NT}][VOL - 0.5\,\Delta t(QPT + QT)]^n = C \quad (9.62)$$

in which $H_b$ is the barometric pressure head, $VOL$ the air volume, and $n$ the polytropic exponent. The exponent $n$ varies from 1 for isothermal to 1.4 for reversible adiabatic. Often, the average value of 1.2 is used. The constant $C$ is evaluated from initial conditions.

Equation 9.62 does not account for heat transfer, but studies by Graze (20) and Martin (36) have shown that it predicts the heads accurately enough for engineering design. For a more exact approach, see Reference 19.

## 9.11   COMPLEX PIPING SYSTEMS

### Series Connection

The discussion thus far has been limited to a single pipe with constant properties and dimensions. A change in pipe properties or wave speed, pipe roughness or diameter, requires a boundary condition because $B$ and $R$ in Eqs. 9.24 and 9.25 depend on them. The solution requires keeping track of the different pipes and variables with proper subscripting. Figure 9.11 shows a pipe junction. The head and discharge at the junction for calculations in pipe $J$ will be designated $HP_{J,NJ}$ and $QP_{J,NJ}$. For pipe $J + 1$, the head and discharge of the same point will be labeled $HP_{J+1,1}$ and $QP_{J+1,1}$. The first subscript refers to the pipe and second to the node. If the loss at the connection is ignored,

$$HP_{J,NJ} = HP_{J+1,1} \quad (9.63)$$

and from continuity

$$QP_{J,NJ} = QP_{(J+1,1)} \quad (9.64)$$

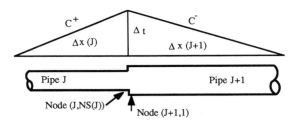

**Fig. 9.11**   Nomenclature for series pipe connection.

The subscripting for constants that apply to each pipe only require one subscript: that is, $N_J$, $B_J$, $R_J$, $A_J$, $\mathbf{a}_J$, etc. The subscript refers to the pipe number.

The $C^+$ and $C^-$ compatibility equations for pipe nodes $(J, NS)$ and $(J + 1, 1)$ are

$$HP_{J,NJ} = CP - B_J QP_{J,NJ} \tag{9.65}$$

$$HP_{(J+1,1)} = CM + B_{(J+1)} QP_{(J+1,1)} \tag{9.66}$$

Solving eqs. 9.63–9.66 gives

$$QP_{J,NS} = \frac{CP - CM}{B_J + B_{J-1}} \tag{9.67}$$

Pipe junctions containing more than two pipes are handled similarly to a series connection. The continuity equation is used and the head in each pipe at the node is the same (assuming no losses). There are more equations, but the method of solving is the same.

A change of pipe material which causes a wave speed change is also handled the same as a series connection. Since the solution must be carried out at the same time in all pipes, $\Delta t$ must be the same, so

$$\Delta t = L_J / [\mathbf{a}_J N_J] \tag{9.68}$$

It will not usually be possible to satisfy Eq. 9.68 without some adjustment in either the lengths or wave speeds. Since the wave speed cannot be determined precisely, allowing 10–15% adjustment in it is permissible. For long valve closure times, the wave speed has little influence on the transient head rise, as demonstrated in Example 9.6. It is therefore desirable to select a set of $N_J$s so that Eq. 9.68 is satisfied without varying any $\mathbf{a}$ by more than about 15%. This is illustrated in Table 9.6, where the number of nodes for three pipes is evaluated with a spreadsheet. The wave speeds and pipe lengths are listed at the top of the table. Conditions are set for the first pipe. The first trial $n_1 = 1$ is used, which gives a $\Delta t$ of 1.129 s. This must be the

**TABLE 9.6  Spreadsheet for Series Pipe**

| a = 3100 | Pipe #1 $L = 3500$ | | | a = 1600 | Pipe #2 $L = 4600$ | | | | a = 2600 | Pipe #3 $L = 8600$ | | |
|---|---|---|---|---|---|---|---|---|---|---|---|---|
| $a_1$ Given | $n_1$ Assumed | $L_1$ Given | $Dt$ Eq. 9.68 | | $n_2$ Assumed | $L_2$ Given | $a_2$ Eq. 9.68 | Conclusion | | $n_3$ Assumed | $L_3$ Given | $a_3$ Eq. 9.68 | Conclusion |
| 3100 | 1 | 3500 | 1.129 | | 1 | 4600 | 4074 | no good | | 2 | 8600 | 3809 | no good |
| 3100 | 1 | 3500 | 1.129 | | 2 | 4600 | 2037 | no good | | 3 | 8600 | 2539 | ok |
| 3100 | 1 | 3500 | 1.129 | | 3 | 4600 | 1358 | no good | | 3 | 8600 | 2539 | ok |
| 3100 | 2 | 3500 | 0.565 | | 5 | 4600 | 1630 | ok | | 6 | 8600 | 2539 | ok |
| 3100 | 4 | 3500 | 0.282 | | 10 | 4600 | 1630 | ok | | 12 | 8600 | 2539 | ok |
| 3100 | 8 | 3500 | 0.141 | | 20 | 4600 | 1630 | ok | | 24 | 8600 | 2539 | ok |

same for all pipes. For pipes 2 and 3 the $n$ values tried are 1 and 2. This produces **a** values of 4074 and 3809, which are too large. Subsequent trial values of $n = 1$, 2, and 3 and 1, 3, and 3 also produces **a** values that are not acceptable. The fourth set of $n_j$ of 2, 5, and 6 produce values that are acceptable. These values or any multiple of them will work.

### Tee Connection

An important principle of transients in pipe systems is that at each pipe junction the transient head is reduced as it is divided into multiple pipes. To demonstrate this, consider a transient head rise of $\Delta H_1$ generated by closing the outlet valve shown in Fig. 9.5. Assume that initially all of the flow is going out the valve so that there is no flow in the pipe past the junction. When the wave arrives at the junction (node $NV$), the head rise in the main pipe will be half of the original transient. This is explained as follows.

Initially, the head tries to increase by the full value $\Delta H_1$ and stop the flow in the upstream pipe. With a head rise, the water in the downstream pipe will accelerate as predicted by $\Delta Q = \Delta H / B$. By continuity, the flow in both sections of pipe must be the same so the $\Delta H$ in both pipes must be the same and equal to $0.5 \, \Delta H_1$.

The $\Delta H_1$ is equally divided as long as the pipe on both sides of the junction is identical. If the downstream pipe is smaller, the head rise will be larger. The limit is when the downstream pipe diameter is zero and the head rise is $\Delta H_1$.

For complex pipe networks, a transient generated at one point does not subject the entire pipe system to high pressures. It only causes high pressures in adjacent pipes because of the division of the wave at junctions.

# 10

# COLUMN SEPARATION AND TRAPPED AIR

## 10.1 COLUMN SEPARATION

If the pressure during a transient drops to vapor pressure, a new boundary condition must be introduced. Consider the case of instant closure of a valve located at the end of a pipe connected to a reservoir. Table 9.1 lists data for such a case (Example 9.2). The diagonal lines in Table 9.1 show how, in the first $L/\mathbf{a}$ second, the high pressure wave travels to the reservoir and stops the flow. In the next $L/\mathbf{a}$ second, the wave is reflected back to the valve at reservoir pressure and the flow reverses. When this wave hits the valve, the pressure drops, first to reservoir pressure; then it tries to drop by another $B$ $\Delta Q = 653.5$ ft to $-603.5$ ft.

Since water cannot sustain pressures below vapor pressure, the water will rupture or vaporize. This is called column separation. The flow rate at the valve will not be zero, as shown in Table 9.1, but will continue flowing into the reservoir with only a small decrease. The change in the flow rate will be $\Delta H/B$ where $\Delta H = H_{NS} - Z_{NS} + H_b - H_{va}$; $H_{NS}$ is the piezometric head at the valve just before the downsurge (in this case it is the reservoir head of 50 ft), $H_b$ is the barometric pressure head, and $H_{va}$ is the absolute vapor pressure head (Table 1.1).

From Example 9.2 and Table 9.1 and assuming $H_b = 30$ ft, $Z = 0$, and $H_{va} = 0.2$ ft, $B = \mathbf{a}/gA = 1500/(32.2 \cdot 0.7854) = 59.3$, $\Delta H = 50 - 0 + 30 - 0.2 = 79.8$ ft, and $\Delta Q = 79.8/59.3 = 1.34$ cfs, so the flow away from the valve after column separation occurs will be 9.67 cfs. This flow away from the valve causes the vapor cavity to grow as predicted by conservation of mass.

The water in the pipe moves toward the reservoir almost as a rigid body. Friction and the pressure difference between the reservoir and the vapor cavity eventually stop and reverse the motion. The time required to stop the flow is much longer than $L/a$.

While the cavity exists, the pressure waves traveling through the pipe rapidly attenuate and the motion becomes rigid-body motion. During the time that the cavity is open, the behavior of the water column can be simulated with surge calculations (see Section 9.9) if desired.

Once the flow reverses, the size of the vapor cavity reduces and the vapor cavity eventually disappears. At the instant the cavity volume goes to zero, a head rise of $aV/g = BQ$ occurs, where $V$ is the liquid velocity just before the cavity disappears. For the next $2L/a$, this high pressure wave travels to the reservoir and back. Depending on its magnitude, it may cause column separation several more times. While the cavity exists, the piezometric head at the valve is vapor pressure.

$$HP_{NS} = Z_{NS} - H_b + H_{va} \tag{10.1}$$

With the cavity located at the valve, the flow is found from the $C^+$ equation from node $N$:

$$QP_{NS} = \{CP - [Z_{NS} - H_b + H_{va}]\}/B \tag{10.2}$$

The size of the vapor cavity ($VOL$) changes in accordance with conservation of mass. The volume change during a time step can be approximated by the average flow at the node times $dt$. The volume at the end of each time step is

$$VOL_{NS} = VOL_{NS} - 0.5dt[QP_{NS} + Q_{NS}] \tag{10.3}$$

when $VOL_{NS}$ becomes zero, the boundary condition returns to the normal condition at a closed valve $QP_{NS} = 0$ and $HP_{NS} = CP$.

Figure 10.1 demonstrates the sequence of events for column separation caused by instant valve closure. The figure also shows computed results that will be discussed in a subsequent section. The $2L/a$ time for this experimental pipeline was about 0.05 s. Figure 10.1 shows that for valve closure at $t = 0$, the head at the valve jumps to about 180 m. It increases slightly due to line packing for $2L/a$ s. It then drops to vapor pressure and stays there until about 0.35 s. Upon collapse of the cavity, another pressure wave is generated at the valve. This pressure rise (and each subsequent rise) is smaller and the $2L/a$ time that it takes to travel to and from the upstream reservoir gets longer. This is because some air is released during column separation and free air reduces the wave speed. The reduced wave speed and attenuation caused by friction reduces the peak pressure. Also note that the time the cavity stays open reduces each cycle. As friction and dissipation

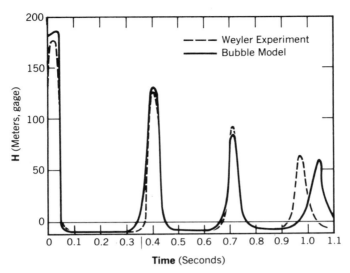

**Fig. 10.1**  Column-separation transient.

caused by vaporization and air release slow down the liquid, column separation will eventually stop and the pressure trace will become a damped sinusoidal shaped wave.

Column separation can also occur at an interior point, especially if the pipe is elevated. Refer again to Table 9.1. If the output were listed for $t > 3$s, the head at each interior node is less than vapor head. The magnitude of the head at these nodes is not accurate because the program is invalid once column separation occurs at the valve. However, the data would show that column separation may occur at the interior points.

Figure 10.2 identifies the variables associated with this boundary condition. Since adjacent nodes may also be experiencing column separation,

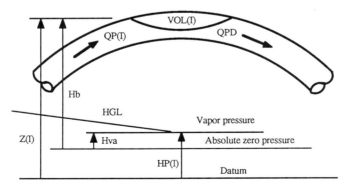

**Fig. 10.2**  Column separation at interior node.

these variables will be used at all nodes. The unknown variables are: $HP$, $QP$, $QPD$, and $VOL$. The same quantities known from the previous time step are $H$, $Q$, $QD$, and $VOL$. The equations are

$$HP_I = Z_{I-}H_b + H_{va} \text{ (during column separation)} \tag{10.4}$$

or

$$HP_I = \frac{(CP + CM)}{2} \text{ (without column separation)} \tag{10.5}$$

$$QP_I = \frac{(CP - HP_I)}{B} \tag{10.6}$$

$$QPD_I = \frac{(HP_I - CM)}{B} \tag{10.7}$$

$$VOL_I = VOL_I + 0.5dt[(QPD_I + QD_I - QP_I - Q_I)] \tag{10.8}$$

The program must check each node to determine when column separation first occurs and select the proper equation for $HP$. It must also check each time step to determine if the cavity is already open and determine when the cavity closes. The logic is shown in the following FORTRAN program steps for interior points.

```
DO 1, I = 2, N
CP = H(I − 1) + B*QD(I − 1) − R*QD(I − 1) ABS(QD(I − 1))
CM = H(I + 1) − B Q(I + 1) + R Q(I + 1)ABS(Q(I + 1))
HP(I) = (CP + CM)/2.
IF ((HP(I) − Z(I) + HB.LE.HVA).OR.(VOL(I).GT.0))THEN
    *COLUNN SEPARATION*
    HP(I) = Z(I) − HB + HVA
    QP(I) = (CP − HP(I))/B
    QPD(I) = (HP(I) − CM)/B
    VOL(I) = VOL(I) + .5 DT*(QPD(I) + QD(I) − QP(I) − Q(I))
     IF(VOL(I).LE.0)THEN !COLLAPSE
        VOL(I) = 0
        HP(I) = (CP + CM)/2.
        QP(I) = (CP − HP(I))/B
        QPD(I) = QP(I)
     END IF
   ELSE
     QP(I) = (CP − HP(I))/B
     QPD(I) = QP(I)
END IF
1 CONTINUE
```

The program steps for column separation at the valve are similar and the logic is the same.

## 10.2   AIR RELEASE DURING COLUMN SEPARATION

The preceding discussion assumes that no air is released during the vaporization process. In almost all cases, some air comes out of solution and will remain as free air when the cavity disappears.

Most liquids contain dissolved air or gas. The saturation pressure is that pressure at which the liquid cannot dissolve any more air. This pressure is often at or slightly below atmospheric pressure. When the pressure drops below saturation, and especially if it drops to vapor pressure and there is some agitation, significant amounts of air can be released. The amount of air released can be estimated if adequate information on the system and the history of the water is available. In most cases, the amount of released air is unknown.

Numerically modeling a system containing free air is difficult. If the air (like the vapor cavity) is assumed to be collected at the computing section, it can be treated as an air-chamber boundary condition. If it is dispersed throughout the pipe, the wave speed will be affected and the numerical techniques discussed in Chapter 9 cannot be used. The dependence of the wave speed on free air was discussed in Section 8.7.

There are at least three methods of handling a variable wave speed. One is the characteristics grid method suggested by Streeter (Section 8.3 of Reference 87). This method appears only to be usable for single straight pipes. Adapting the method to more complex pipe systems appears doubtful. The second and most common method is using interpolations (Section 8.4 of Reference 87). This method has inherent difficulties associated with numerical dampening and a requirement to use a predictor–corrector type of solution. However, it is a viable way of handling the variable wave speed. The third method assumes that all the air is collected at computing nodes and handles it like an air-chamber boundary condition. Since the development of Eq. 8.13 was independent of how the air was distributed, the overall wave speed is the same regardless of whether the air is uniformly distributed or collected at a few points.

Each method has limitations and none is universally accepted. The biggest problem is that none of the present methods adequately model the dissipation caused by the air. It will be some time before this problem is resolved.

The third method will be used to demonstrate what influence air has on transients by use of an example. The example is selected from a project involving field test and numerical simulation of a 24-in. diameter 17,000-ft long concrete pipeline. This pipeline experienced frequent breaks over many months of operation. A computer simulation of the system identified that the transient pressures responsible for the pipe breakage were being caused by rapid valve closures. Field tests were made to verify the results of the computations.

Some of the field tests consisted of establishing a small flow through one of the valves and generating a transient by instantly closing the valve. Results of one such test is shown in Fig. 10.3. Instead of producing a square wave, typical of transients caused by instant valve closures, a damped sinusoidal-shaped wave was generated. The wave speed for the pipe, as calculated from Eq. 8.11, was 3640 fps. The wave speed calculated from the field data shown in Fig. 10.3 was 1170 fps. This large reduction in wave speed from its theoretical value suggested trapped air in the pipeline. The air was likely trapped in the pipe bell and spigot connections and at two high points where there were no air release valves. This mass of air was estimated and then used with Eq. 8.13 to calculate the wave speed. The results agreed very closely with the field tests.

This amount of air was then incorporated into a computer program that concentrated the air at computing points. The air was handled like a normal air-chamber boundary condition. The air was allowed to expand and contract with varying pressure, but was not allowed to move through the pipe. Details of the development of the computer program are included in Reference 66. The computer simulation of one field test is shown in Fig. 10.3. Comparison of the measured and computed lines show gratifying similarities. The magnitudes, shapes, and timing of the two transients are quite similar.

The only two items required in producing the computed transient shown in Fig. 10.3 are the volume of trapped air, which was estimated from analyzing the pipeline, and the amount of resistance of the flow in and out of the air chamber.

The first computer program used to simulate the effect of entrapped air allowed the air pockets to expand and contract with no resistance. This produced oscillations in the solution caused by reflections from air pockets. Dampening was introduced by assuming a small resistance of the flow in and out of the air chamber. This has a similar effect to that created by the

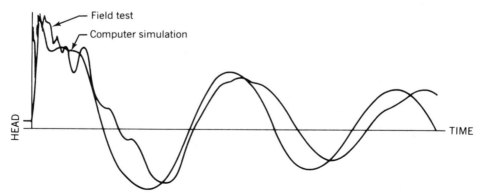

**Fig. 10.3**    Field test and computer results of air in a pipe.

interpolations used by other methods. It is felt that the energy dissipation caused by this small resistance represents a physical phenomena and is not just an arbitrary mathematical dampening technique. For example, much of the air will be entrapped in the pipeline as either individual tiny bubbles or as small pockets trapped in crevices of the boundary or couplings. Surface tension creates a pressure differential between the air inside the bubble and the water outside. This pressure differential is on the order of a fraction of a psi. This is the same order of magnitude of pressure differential required between the water and the air chamber in order to produce a required numerical dampening. Additional research is required to relate the amount of dampening physically needed to simulate the physics of the system.

It is instructive to compare the results of the model that includes the trapped air with results of the simple waterhammer program (Fig. 9.3). The field test and computer simulation shown in Fig. 10.3 was for instant closure of a valve at the end of a pipe. The initial velocity was low, so friction was negligible. The program listed in Fig. 9.3 would produce a square wave with little attenuation for such a case.

Figure 10.4 is the computer simulation of the same test shown in Fig. 10.3 for instant valve closure with the calculated wave speed of 3640 fps and the field measured wave speed of 1170 fps. In both cases, a square wave is produced. The difference is in the period and magnitude of the waves. In comparing the results of Figs. 10.3 and 10.4, two obvious differences are noticeable. First, the shape of the waves is totally different, and second, there is virtually no attenuation for the computer results shown in Fig. 10.4. This points out the profound effect that air has on the transient. It not only reduces the wave speed and the pressure rise, but causes a major change in the shape and the attenuation of the transient. Using the basic waterhammer program with the computed wave speed (3600 fps) predicts a pressure rise

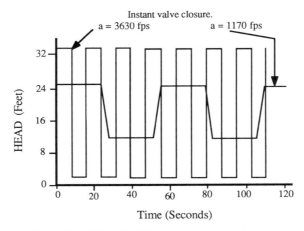

**Fig. 10.4**  Transient for instant valve closure.

that is three times higher than was measured in the field for instant closure. Using a 1170-fps wave speed produced the proper initial head rise and the correct period, but the shape and attenuation were totally incorrect.


## 10.3  TRAPPED AIR

Air trapped as large pockets in pipelines can lead to significantly higher transients than if there was no air. Air trapped at high pressure stores large quantities of energy. If the air is forced to move from one location to another, it can cause local accelerations resulting in high velocities and subsequently high transient pressures. Analytical work (36, 87) has shown that these pressures can be many times the system pressure.

One example of such a transient is when large air pockets are allowed to go through a valve, orifice, or other type of control device. Consider a system where water containing trapped air is flowing through a valve. When a large air pocket passes through the valve, the air flows through much more easily than water, since it is less dense. The pressure drop across the valve suddenly reduces, causing the pressure just upstream from the valve to drop, and the upstream water column is allowed to accelerate. Once all the air passes through the valve, the water is caused to suddenly decelerated back to its steady-state velocity. This sudden reduction in velocity causes a transient pressure rise.

This problem is made worse when there are multiple large air pockets in the pipe. The pockets divide the system into smaller masses of water that can be accelerated to higher velocities. The air stores energy, which maintains a high acceleration force. The best solution to this problem is to prevent large air pockets from forming. Since this is not always possible, air release valves should be placed to remove the air slowly before it reaches a control valve or orifice.

Another source of transients caused by large quantities of trapped air occurs if compressed air is released through a large air valve. Since the air has a low density, it flows at high rates through the valve (usually at sonic velocity). This reduces the pressure inside the pipe and allows the water column to accelerate. Once the air is expelled, the air valve closes and a transient is generated.

There are methods of controlling this type of transient. First and most important is to fill slowly. Second, install large enough air valves that the air is released without pressurizing. Third, lay the pipeline to grade and install air valves at high points. If the pipe is horizontal or on a flat grade, install air valves at regular intervals. Fourth, flush the system at moderate velocities (2–4 fps) and low pressure to move the air to the air valves. Fifth, be sure that there are air valves upstream from control valves so the air is removed from the pipe (87) and does not pass through the control valve. Sixth, install both air-vacuum valves to take care of initial filling or draining and air-release valves that can continually release air accumulating in the pipe. There is additional discussion of air valves in Section 4.6.

Another source of air-related transients is filling a pipeline containing large pockets of air. Similar problems occur at pump start-up or at opening of upstream control valves when the line is presumably full but actually contains trapped air. Figure 10.5, from Reference 87, shows the magnitude of the transient created by opening an upstream valve when there is trapped air at the downstream end of the pipe. When the valve shown in Fig. 10.5 is opened, the water accelerates rapidly. It attains a velocity of almost 25 fps before the air pocket decelerates and eventually reverses the flow. For this example, the maximum head in the pipe is almost eight times the reservoir head.

Martin presented parametric curves for predicting the head rise caused by rapidly filling pipes containing trapped air (36). The system is the same as that shown in Fig. 10.5. Martin's analysis showed that the maximum absolute head $H^*_{max}$ depends only on three parameters: $H^*_o/H_b$, $fv_o/d^3$ and $n$, in which $H^*_o$ is the absolute reservoir head, $H_b$ is the barometric pressure head, $f$ the friction factor, $v_o$ the initial air volume, $d$ the pipe diameter, and $n$ the polytropic exponent. Figure 10.6 shows this relationship for $n = 1.2$. The data assume instant valve opening.

**Example 10.1.** Use Fig. 10.6 to check the predictions for the example shown in Fig. 10.5.

The system data are $v_o = 0.7854 \cdot 3.5^2 \cdot 40 = 385$ ft$^3$, $H_o = 100$ ft and assume $H_b = 32$ ft, $f = 0.02$, and $d = 3.5$ ft.

$$fv_o/d^3 = 0.18, \quad H^*_o/H_b = 132/32 = 4.12$$

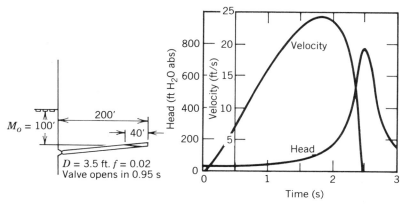

**Fig. 10.5**  Transient caused by filling pipe (87). (Courtesy McGraw-Hill Book Co.)

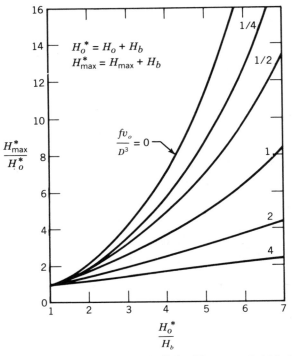

**Fig. 10.6**  Transient caused by trapped air (36). (Courtesy British Hydromechanics Research Association and C. S. Martin.)

From Fig. 10.6, $H^*_{max}/H^*_o = 6.1$, and from Fig 10.5, $H^*_{max} = 780$ ft. So,

$$H^*_{max}/H^*_o = 780/132 = 5.91$$

The value of $H_b$ for the problem in Fig. 10.5 was not given, but using $H_b = 32$ ft is a reasonable value. With this value of $H_b$, the two methods give comparable answers.

For the confined system with instant opening of the valve and no air release from the pipe, the location of the air pocket has no influence on the maximum pressure. When there is air release, the pressures are higher when the air pocket is close to the upstream reservoir. For a confined system, the head rise would also be independent of whether the air is in one large pocket or distributed in a number of smaller pockets.

***Example 10.2.*** Consider an 18-in., 1200-ft long asbestos cement pipeline used for transporting water from a storage tank at elevation 4100 ft to a second storage tank at elevation 3900 ft. Assume the line was shut down for minor repairs and when refilled, some air was trapped at a high point at elevation 3950 ft. The initial air volume $v$ is 100 ft$^3$ at atmospheric pressure. The line is to be pressurized by opening a valve at the upstream storage tank. Calculate the potential surge caused by compressing the air.

For the system: $f = 0.014$, $H_b = 29.2$ ft (Eq. 1.11), $d = 1.5$ ft, $v_o = 100$ ft$^3$, $H_o^* = 4100 - 3950 + 29.2 = 179$ ft. (The air is assumed to be at atmospheric pressure.)

$$\frac{fv_o}{d^3} = 0.41, \quad \frac{H_o^*}{H_b} = \frac{179}{29.2} = 6.14$$

From Fig 10.6, $H_{max}^*/H_o^* = 10.2$, so $H_{max}^* = 10.2\,(179) = 1827$ ft (abs)

The maximum gauge pressure will be $(1827 - 29.2)/2.31 = 778$ psi. This is far above the design pressure and even above the hydrostatic test pressure of the pipe (see Chapter 2). This could easily rupture the pipe.

This calculated peak pressure is only accurate if the filling valve is opened in less time than it takes to accelerate the water in the pipe. For this problem, the time for the water to reach its maximum velocity is on the order of 30 s.

When valves are installed in long pipelines, they only control the flow at very small openings (see Examples 4.1 and 8.2). For this pipeline, once the valve is open about 25%, it offers little resistance to the flow. As a result, a total opening time of about 2 min will effectively be the same as a 30-s opening.

The analysis shows that if the filling valve at the upstream reservoir is opened in about 2 min or less, the pressure in the pipe can be as high as 778 psi.

# APPENDIX A
# CONVERSION FACTORS

*Length*

in. = 2.54 cm
ft = 0.3048 m
mi. = 5,280 ft

*Volume*

U.S. gal = 3.785 L
Acre-Ft(A-F) = 43,560 cu-ft
Million gal/day (mgd) = 3.0689 A-F
A-F = 1,233 cu-m
1 cu-ft = 7.48 U.S. gal

*Flow*

cfs = 448.8 gpm (U.S. gal/min)
cfs = 0.0283 m$^3$/s
cfs = 1.984 A-F in 24 h.
mgd = 1.547 cfs

*Pressure* (water at 50°F)

psi = 2.308 ft
psi = 6.895 kPa
in. HG = 1.133-ft water

*Miscellaneous*

kW = 1.341 horsepower

# REFERENCES

1. ANSI/AWWA C403-78, *Standard Practice for the Selection of Asbestos-Cement Pipe*, American Water Works Association, Denver, CO.

2. ASCE, "Design of Water Intake Structures for Fish Protection," The Task on Fish-Handling of Intake Structures of the Committee of Hydraulic Structures, American Society of Civil Engineers, New York, 1982.

3. ASCE, "Pipeline Design for Water and Wastewater," Pipeline Division, American Society of Civil Engineers, New York, 1975, 128 pp.

4. Baasiri, M., and Tullis, J. P., "Air release during column separation," *J. Fluids Eng.*, **105**, 1983, pp. 113–118.

5. Ball, J. W., "Sudden enlargements in pipelines," *J. Power Div.*, *ASCE*, **98**, (PO4), Proc. Paper 3340, December 1962, pp. 15–27.

6. Ball, J. W., and Sweeney, C. E., "Incipient cavitation damage in sudden enlargement energy dissipators," *Proceedings of the Conference of Cavitation*, Edinburgh, Scotland, September 3–5, 1974, Institute of Mechanical Engineers, London, 1974, pp. 73–79.

7. Ball, J. W., Tullis, J. P., and Stripling, T., "Predicting cavitation in sudden enlargements," *J. Hydraul. Div. ASCE*, **101**(HY7), July 1975, pp. 857–870.

8. Baquero, F., "Cavitation damage in elbows," Masters Thesis, Colorado State University, Fort Collins, CO, 1977.

9. Chincholle, L., and Guymord, D., "Detection de la cavitation erosive et de l'abrasion," *IAHR Symposium*, 1980, Tokyo, pp. 117–130.

10. Clyde, E. S., "Cavitation scale effects in pipe elbows," Masters Thesis, Colorado State University, Fort Collins, CO, 1977.

11. Clyde, E. S., and Tullis, J. P., "Aeration scale effects," *ASCE Specialty Conference*, Boston, MA, American Society of Civil Engineers, New York, August 1983.

12. Colebrook, C. F., "Turbulent flow in pipes, with particular reference to the transition region between the smooth and rough pipe laws," *J. Inst. Civ. Eng. Lond.*, **11**, 1938–1939, pp. 133–156.

13. Crane Co., "Flow of Fluids Through Valves, Fittings, and Pipe," Technical Paper No. 410, Engineering Division, Crane Company, New York, 1979.

14. Hardy, Cross, "Analysis of Flow in Networks of Conduits or Conductors," University of Illnois Bulletin 286, Urbana, IL, November 1936.

14a. Davis, C.V., and Sorensen, K.E., *Handbook of Applied Hydraulics*, Third edition, McGraw-Hill, New York, 1969.

15. Davis, R. T., "Aerating Butterfly Valves to Suppress Cavitation," Masters thesis Utah State University, Logan, Utah 1986.

16. Deeprose, W. M., King, N. W., McNulty, P. J., and Pearsall, I. S., "Cavitation noise, flow noise and erosion," *Proceedings of the Conference on Cavitation*, Edinburgh, Scotland, September 3–5, 1974, Institute of Mechanical Engineers, London, 1974, pp. 373–381.

17. Evans, W. E., and Crawford, C. C., "Design charts for air chambers on pumped lines," *Trans. Am. Soc. Civ. Eng.*, **119**, 1954, pp. 1025–1045.

18. Fang, K. S., and Kooslhof, F., "Determination of NPSH on large centrifugal pumps and Thoma's law of similarity," *Cavitation Forum*, Doctorate American Society of Mechanical Engineers, New York, 1971.

19. Govindarajan, R., "Cavitation size scale effects," Thesis Colorado State University, Fort Collins, CO, 1972.

20. Graze, H. R., "A rational thermodynamic equation for air chamber design," *Third Australasian Conference on Hydraulics and Fluid Mechanics*, November 25–29, 1968, pp. 57–61.

21. Grist, E., "Net positive suction head requirements for avoidance of unacceptable cavitation erosion in centrifugal pumps," *Proceedings of the Conference on Cavitation*, Edinburgh, Scotland, September 3–5, 1974, Institute of Mechanical Engineers, London, 1974.

22. Haltiner, G. J., and Martin, F. L., *Dynamical and Physical Meteorology*, McGraw-Hill, New York, 1957.

23. Hammitt, F. G., "Cavitation damage scale effects—state of art summarization," *J. Hydraul. Res. Int. Assoc. Hydraul. Res.*, **13**(1), 1975, pp. 1–17.

24. Hammitt, F. G., "Effect of gas content upon cavitation inception, performance, and damage," *J. Hydraul. Res. Int. Assoc. Hydraul. Res.*, **10**(3), 1972, pp. 259–290.

25. *Hydraulic Institute Standards*, 12th edition, New York.

26. ISA S39.1 and .2, *Control Valve Capacity Test Procedure for Incompressible Fluids*, Instrument Society of American Standards. Pittsburgh, Pa.

27. James, L. D., and Lee, R. R., *Economics of Water Resources Planning*, McGraw-Hill, New York, 1971, 615 pp.

28. Jeppson, R., *Analysis of Flow in Pipe Networks*, Ann Arbor Science, Ann Arbor, MI, 1976, 164 pp.

29. Kallas, D. H., and Lechtman, J. Z., "Cavitation erosion," in C. V. Rosato and R. T. Schwartz, eds., *Environmental Effects on Polymeric Materials*, C. V.

Rosato and R. T. Schwartz (eds.), Wiley-Interscience, New York, 1968, pp. 223–280.

30. Karassik, I. J., Krutzcsh, W. C., Fraser, W. H., and Messina, J. P., *Pump Handbook*, McGraw-Hill, New York, 1976.

31. King, H. W., and Brater, E. F., *Handbook of Hydraulics*, McGraw-Hill, New York, 1970, 578 pp.

32. Knapp, R. T., "Recent investigations of the mechanics of cavitation and cavitation damage," *Trans. Am. Soc. Mech. Eng.*, **77**, October 1955, pp. 1045–1054.

33. Knapp, R. T., Daily, J. W., and Hammitt, F. G., *Cavitation*, McGraw-Hill, New York, 1970, 578 pp.

34. Lichtarowicz, A., and Pearce, I. D., "Cavitation and aeration effects in long orifices," *Proceedings of the Conference on Cavitation*, Edinburgh, Scotland, September 3–5, 1974, Institute of Mechanical Engineers, London, 1974, pp. 129–144.

35. Manning, R., "Flow of water in open channels and pipes," *Trans. Inst. Civ. Eng., Ireland*, **20**, 1890.

36. Martin, C. S., "Entrapped air in pipelines," *Proceedings of the Second International Conference on Pressure Surges*, London, September 22–24, 1976, BHRA Fluid Engineering, Cranfield, Bedford, England, 1976.

37. Miller, D. S., "Internal Flow, A Guide to Losses in Pipe and Duct Systems," British Hydromechanics Research Association, Cranfield, Bedford, England.

38. Moody, L. F., "Friction factors for pipe flow," *Trans. Am. Soc. Mech. Eng.*, November 1944, **66**, pp. 671–684.

39. Morrison, E. B., "Nomograph for the design of thrust blocks," *Civ. Eng.*, **39**, June 1969, pp. 50–51.

40. Mumford, B. L, "Cavitation limits and the effect of aeration on cone valves," Masters Thesis, Utah State University, Logan, UT, 1985.

41. Nikuradse, J., "Gesetzmassigkeiten der turbulenten Strömung in glatten Rohren," *Verh. Dtsch. Ing. Forschungsheft*, **356**, 1932.

42. Numachi, F., Yamabe, M., and Oba, R., "Cavitation effect on the discharge coefficient of the sharp-edged orifice plate," *J. Basic Eng.*, **82E**(1), March, 1960, pp. 1–11.

43. O'Brien, T., "Needle valves with abrupt enlargements for the control of a high head pipeline – model studies and hydraulic design," *J. Inst. of Eng. Aust.* October–November, 1966, pp. 141–149.

44. Provoost, G. A., "The dynamic characteristic of non-return valves," *Proceedings, IAHR 11th Symposium on Operating Problems of Pump Stations and Power Plants*, Amsterdam, September 1982.

45. Rahmeyer, W. J., "Predicting Hydrodynamic Noise From Cavitating Valves," American Society of Mechanical Engineers, Pressure Vessel's and Piping Conferences, Pittsburg Pa, June 1988, 63 pp.

46. Rahmeyer, W. J., "Predicting and modeling cavitation damage: sudden enlargement," Doctorate Thesis, Colorado State University, Fort Collins, CO, 1980.

47. Rahmeyer, W. J., and Tullis, J. P., "Cavitation limits for butterfly valves,"

*Symposium on Cavitation Erosion in Fluid Systems*, American Society of Mechanical Engineers, New York, 1981.

48. Rouse, H., *Elementary Mechanics of Fluids*, John Wiley, New York, 1960.
49. Rouse, H., and Jezdinsky, V., "Cavitation and Energy Dissipation in Conduit Expansion," *Int. Assoc. Hydraul. Res.*, **1**, Paper 1.28, 1965, pp. 1–8.
50. Rouse, H. A., and Jezdinsky, V., "Cavitation and energy dissipation in conduit expansion," *Proceedings of the 11th Congress of International Association for Hydraulics Research*, Leningrad, USSR, 1965, pp. 1–4.
51. Russell, S. O., and Ball, J. W., "Sudden enlargement energy dissipator for mica dam," *J. Hydraul. Div.*, ASCE, **93**(HY4), Proc. Paper 5337, July 1967, pp. 41–56.
52. Sarpkaya, T., "Torque and cavitation characteristics of butterfly valves," *J. Appl. Mech.*, **28**, **29**(E4), December, 1961, pp. 511–606.
53. Smith, P. E., "Supercavitation in pipe flow," Masters Thesis, Colorado State University, Fort Collins, CO, 1976.
54. Stark, R. M., and Nichols, R. L., *Mathematical Foundations for Design: Civil Engineering Systems*, McGraw-Hill, New York, 1972, 566 pp.
55. Stephenson, D., *Pipeline Design for Water Engineers*, 2nd edition, Elsevier, New York, 1981.
56. Stierlin, K., "Pressure reducing plant for a water power station," *Escher Wyss News*, **30**(1), 1957, pp. 19–28.
57. Stiles, G. F., "Cavitational Tendencies of Control Valves for Paper Pulp Service," Preprint No. 17.1-3-66, Preprint Instrument Society of America, 12 pp.
58. Streeter, V. L., and Wylie, E. B., *Fluid Mechanics*, McGraw-Hill, New York, 1975, 752 pp.
59. Stripling, T. C., "Cavitation damage scale effects: sudden enlargements," Doctorate Thesis, Colorado State University, Fort Collins, CO, 1975.
60. Sweeney, C. E., "Cavitation damage in sudden enlargements," Thesis presented to Colorado State University, Fort Collins, CO, 1974, 81 pp.
61. Sweeney, C. E., and Rockwell, G. E., "Pump sump design acceptance through hydraulic model testing," *Proceedings, IAHR 11th Symposium on Operating Problems of Pump Stations and Power Plants*, Amsterdam, September 1982.
62. Tullis, J. P., "Cavitation scale effects for valves," *J. Hydraul. Div.*, ASCE, **99**(HY7), Proc. Paper 9874, July 1973, pp. 1109–1128.
63. Tullis, J. P., "Choking and supercavitating valves," *J. Hydraul. Div.*, ASCE, **97**(HY12) Proc. Paper 8593, December 1971, pp. 1931–1945.
64. Tullis, J. P., "Comparison of Torque and $C_v$ Characteristics of Several Leaf Designs for a 24-Inch Butterfly Valve," Hydraulics Program Report No. 29, Utah Water Research Laboratory, Utah State University, Logan, UT, January 1980, 46 pp.
65. Tullis, J. P., "Friction Factor Tests on Spiral-Rib Pipe," Hydraulics Program Report No. 83, Utah Water Research Laboratory, Utah State University, Logan, UT, April 1983.
66. Tullis, J. P., "Influence of Entrapped Air on Transients," International Institute on Hydraulic Transients and Cavitation, São Paulo, Brazil, July 12–30, 1982, pp. C3.1–C3.17.

67. Tullis, J. P., "Modeling Cavitation for closed conduit flow," *J. Hydraul. Div.*, *ASCE*, **107**(HY11), Proc. Paper 16650, November 1981, pp. 1335–1349.
68. Tullis, J. P., "Modeling in design of pumping pits," *J. Hydraul. Div.*, *ASCE*, **105**(HY9), Proc. Paper 14812, September 1979, pp. 1053–1063.
69. Tullis, J. P. "Testing valves for cavitation," *Proceedings of the Conference on Cavitation*, Edinburgh, Scotland, September 3–5, 1974, Institute of Mechanical Engineers, London, 1974.
70. Tullis, J. P., and Marschner, B. W., "A review of cavitation research on valves," *J. Hydraul. Div.*, *ASCE*, **94**(HY1), Proc. Paper 5705, January 1968, pp. 1–16.
71. Tullis, J. P., Power III, J. J., Shiers, P. F., and Hall, W. W., "Perforated plates as hydraulic energy dissipators," *Hydraulics Specialty Conference*, ASCE Chicago, IL, 1980.
72. Tullis, J. P., and Skinner, M. M., "Reducing cavitation in valves," *J. Hydraul. Div.*, *ASCE*, **94**(HY6), Proc. Paper 6255, November 1968, pp. 1475–1488.
73. Tullis, J. P., and South, W. D., "Cavitation and Discharge Calibration of Media Butterfly Valves," Hydraulics Program Report No. 18, Utah Water Research Laboratory, Utah State University, Logan, UT, April 1979.
74. Tullis, J. P., and South, W. D., "Hydraulic Tests on CLA-VAL Company Hytrol Valves," Hydraulics Program Report No. 3, Utah Water Research Laboratory, Utah State University, Logan, UT, March 1978.
75. Tullis, J. P., Streeter, V. L., and Wylie, E. B., "Waterhammer analysis with air release," *Proceedings of the 2nd International Conference on Pressure Surges*, London, September 22–24, 1976, BHRA Fluid Engineering, Cranfield, Bedford, England, 1976, pp. C3.35–C3.48.
76. U.S. Department of Interior, "Friction Factors for Large Conduits Flowing Full," Water Resources Technical Publication, Engineering Monograph No. 7, U.S. Government Printing Office, Washington, DC, 1977.
77. Walker, R., "Pump Selection," Ann Arbor Science, Ann Arbor, MI, 1972.
78. Wang, J.-S., and Tullis, J. P., "Turbulent flow in the entry region of a rough pipe," *J. Fluids Eng.*, *Trans. of the ASME*, March 1974, pp. 62–68.
79. Watkins, R. K., "Principles of Structural Performance of Buried Pipes," Utah State University, Logan, UT.
80. Watson, W. W., "Evolution of multijet sleeve valve," *J. Hydraul. Div.*, *ASCE*, **103**(HY6), June 1977, pp. 617–631.
81. Watters, G. Z., *Modern Analysis and Control of Unsteady Flow in Pipelines*, Ann Arbor Science, Ann Arbor, MI, 1979, 251 pp.
82. Whittington, N. C., "Cavitation scale effects for generalized valve shapes," Masters Thesis Colorado State University, Fort Collins, CO, 1970.
83. Winn, W. P., and Johnson, D. E., "Cavitation parameters for outlet valves," *J. Hydraul. Div. ASCE*, **96**(HY12), Proc. Paper 7771, December 1970, pp. 2519–2533.
84. Wislicenus, G. F., *Fluid Mechanics of Turbo-machinery*, Dover Publications, New York, 1965.
85. Wood, D. J., "An explicit friction factor relationship," *Civ. Eng.*, **36**, December 1966, pp. 60–61.

86. Wood, D. J., and Charles, C. O. A., "Hydraulic network analysis using linear theory," *J. Hydraul. Div.*, *ASCE*, **98**(HY7), July 1972, pp. 1157–1170.

87. Wylie, E. B., and Streeter, V. L., *Fluid Transients*, McGraw-Hill, New York, 1978, 384 pp.

88. Yanshin, B. N., *Hydrodynamic Characteristics of Valves and Pipeline Components*, Mashinostroenie, Moscow, 1965, 26 pp.

89. Yarabeck, R. R., "Hydraulic problems in small transmission systems," *Proceedings of the Institute of Control of Flow in Closed Conduits*, Colorado State University, Fort Collins, CO, August 9–14, 1970, pp. 31–44.

# INDEX